GM Diesel Engine Owners Workshop Manual

Matthew Minter

Models covered
This manual covers the GM 1598 cc & 1699 cc Diesel engines used in the Vauxhall Astra, Belmont & Cavalier, Opel Astra, Kadett, Ascona & Vectra, and Bedford/Vauxhall Astra Vans

Does not cover turbocharged engines, or 1686 cc normally-aspirated engine

(1222-7W7)

ABCDE
FGHIJ
KLMNO
PQ
2

Haynes Publishing
Sparkford Nr Yeovil
Somerset BA22 7JJ England

Haynes Publications, Inc
861 Lawrence Drive
Newbury Park
California 91320 USA

Acknowledgements

Thanks are due to Vauxhall Motors Limited for the supply of technical information, and for the loan of the vehicle used in the preparation of this book. Champion Spark Plug provided replacement component information, while Duckhams Oils provided lubrication data. Certain illustrations are the copyright of Vauxhall Motors Limited, and are used with their permission. Thanks are also due to Sykes-Pickavant who provided some of the workshop tools, and to all those people at Sparkford who helped in the production of this manual.

© Haynes Publishing 1994

A book in the **Haynes Owners Workshop Manual Series**

Printed by J. H. Haynes & Co. Ltd., Sparkford, Nr Yeovil, Somerset BA22 7JJ, England

All rights reserved. No part of this book may be reproduced or transmitted in any form or by any means, electronic or mechanical, including photocopying, recording or by any information storage or retrieval system, without permission in writing from the copyright holder.

ISBN 1 85010 992 3

British Library Cataloguing in Publication Data
A catalogue record for this book is available from the British Library.

We take great pride in the accuracy of information given in this manual, but vehicle manufacturers make alterations and design changes during the production run of a particular vehicle of which they do not inform us. No liability can be accepted by the authors or publishers for loss, damage or injury caused by any errors in, or omissions from, the information given.

Restoring and Preserving our Motoring Heritage

Few people can have had the luck to realise their dreams to quite the same extent and in such a remarkable fashion as John Haynes, Founder and Chairman of the Haynes Publishing Group.

Since 1965 his unique approach to workshop manual publishing has proved so successful that millions of Haynes Manuals are now sold every year throughout the world, covering literally thousands of different makes and models of cars, vans and motorcycles.

A continuing passion for cars and motoring led to the founding in 1985 of a Charitable Trust dedicated to the restoration and preservation of our motoring heritage. To inaugurate the new Museum, John Haynes donated virtually his entire private collection of 52 cars.

Now with an unrivalled international collection of over 210 veteran, vintage and classic cars and motorcycles, the Haynes Motor Museum in Somerset is well on the way to becoming one of the most interesting Motor Museums in the world.

A 70 seat video cinema, a cafe and an extensive motoring bookshop, together with a specially constructed one kilometre motor circuit, make a visit to the Haynes Motor Museum a truly unforgettable experience.

Every vehicle in the museum is preserved in as near as possible mint condition and each car is run every six months on the motor circuit.

Enjoy the picnic area set amongst the rolling Somerset hills. Peer through the William Morris workshop windows at cars being restored, and browse through the extensive displays of fascinating motoring memorabilia.

From the 1903 Oldsmobile through such classics as an MG Midget to the mighty 'E' type Jaguar, Lamborghini, Ferrari Berlinetta Boxer, and Graham Hill's Lola Cosworth, there is something for everyone, young and old alike, at this Somerset Museum.

Haynes Motor Museum

Situated mid-way between London and Penzance, the Haynes Motor Museum is located just off the A303 at Sparkford, Somerset (home of the Haynes Manual) and is open to the public 7 days a week all year round, except Christmas Day and Boxing Day.

Telephone 01963 440804.

Contents

	Page
Acknowledgements	2
About this manual	4
Introduction to the GM 16D & 16DA engine	4
Weights and capacities *(also see Chapter 8, page 104)*	5
Buying spare parts	6
General repair procedures	7
Tools and working facilities	8
Safety first!	10
Routine maintenance *(also see Chapter 8, page 104)*	11
Recommended lubricants and fluids	15
Fault diagnosis	16
Chapter 1 Engine *(also see Chapter 8, page 104)*	19
Chapter 2 Cooling system *(also see Chapter 8, page 104)*	63
Chapter 3 Fuel and exhaust systems *(also see Chapter 8, page 104)*	70
Chapter 4 Clutch and transmission	85
Chapter 5 Braking system	86
Chapter 6 Suspension and steering *(also see Chapter 8, page 104)*	88
Chapter 7 Electrical system	90
Chapter 8 Supplement: Revisions and information on later models	104
Conversion factors	126
Index	127

About this manual

Its aim

The aim of this manual is to help you get the best value from your vehicle. It can do so in several ways. It can help you decide what work must be done (even should you choose to get it done by a garage), provide information on routine maintenance and servicing, and give a logical course of action and diagnosis when random faults occur. However, it is hoped that you will use the manual by tackling the work yourself. On simpler jobs it may even be quicker than booking the car into a garage and going there twice, to leave and collect it. Perhaps most important, a lot of money can be saved by avoiding the costs a garage must charge to cover its labour and overheads.

The manual has drawings and descriptions to show the function of the various components so that their layout can be understood. Then the tasks are described and photographed in a step-by-step sequence so that even a novice can do the work.

Unlike most Haynes manuals, which cover a particular vehicle in different trim levels and engine sizes, this book covers one engine and its associated equipment as fitted to a range of vehicles. Items which are common to Diesel and petrol models – eg bodywork, transmission and running gear – are not covered in this book.

Its arrangement

The manual is divided into Chapters, each covering a logical sub-division of the vehicle. The Chapters are each divided into Sections, numbered with single figures, eg 5; the Sections are divided into paragraphs, or into sub-sections and paragraphs.

It is freely illustrated, especially in those parts where there is a detailed sequence of operations to be carried out. There are two forms of illustration: figures and photographs. The figures are numbered in sequence with decimal numbers, according to their position in the Chapter – eg Fig. 6.4 is the fourth drawing/illustration in Chapter 6. Photographs carry the same number (either individually or in related groups) as the Section and paragraph to which they relate.

There is an alphabetical index at the back of the manual as well as a contents list at the front. Each Chapter is also preceded by its own individual contents list.

References to the 'left' or 'right' of the vehicle are in the sense of a person in the driver's seat facing forwards.

Unless otherwise stated, nuts and bolts are removed by turning anti-clockwise, and tightened by turning clockwise.

Vehicle manufacturers continually make changes to specifications and recommendations, and these, when notified, are incorporated into our manuals at the earliest opportunity.

We take great pride in the accuracy of information given in this manual, but vehicle manufacturers make alterations and design changes during the production run of a particular vehicle of which they do not inform us. No liability can be accepted by the authors or publishers for loss, damage or injury caused by any errors in, or omissions from, the information given.

Introduction to the GM 16D & 16DA engine

This small Diesel engine, intended for passenger and light commercial vehicles, was introduced to the UK in mid-1982. It is derived from the 16S petrol engine, which it closely resembles in many respects. Like the 16S it is available in the Vauxhall Astra, Belmont and Cavalier, the Opel Kadett and Ascona, and their commercial equivalents (sold in the UK under the Bedford label).

Although on paper the 16D/16DA engine may appear to lack power when compared with its petrol relatives, on the road it copes well with all conditions up to maximum legal speeds. The slight trade-off in performance is potentially more than repaid by gains in fuel economy and engine longevity, and ultimately by a higher resale value.

The DIY mechanic will find routine maintenance straightforward, with good access to the commonly required items. Maintenance of the engine itself is in fact minimal, and apart from more frequent oil changing there is considerably less to do than on the petrol equivalent. Investment in one or two special tools will be necessary for some major tasks.

Weights and capacities

For modifications, and information applicable to later models, see Supplement at end of manual

The figures are representative; consult individual vehicle documents for confirmation.

Kerb weights*
Astra/Kadett:
- Three-door Hatchback .... 930 to 955 kg (2051 to 2106 lb)
- Five-door Hatchback .... 940 to 975 kg (2072 to 2150 lb)
- Three-door Estate .... 990 kg (2183 lb)
- Five-door Estate .... 998 to 1010 kg (2201 to 2227 lb)
- Van .... 1045 kg (2304 lb)

Belmont .... 958 to 967 kg (2112 to 2132 lb)

Cavalier/Ascona:
- Four-door Saloon .... 1045 kg (2304 lb)
- Hatchback .... 1070 kg (2359 lb)
- Estate .... 1115 kg (2458 lb)

Gross vehicle weights*
Astra/Kadett:
- Three-door Hatchback .... 1435 to 1450 kg (3164 to 3197 lb)
- Five-door Hatchback .... 1445 to 1470 kg (3185 to 3241 lb)
- Five-door Estate .... 1500 to 1580 kg (3306 to 3484 lb)
- Van .... 1580 kg (3484 lb)

Belmont .... 1405 to 1445 kg (3097 to 3186 lb)

Cavalier/Ascona:
- Saloon .... 1545 kg (3406 lb)
- Hatchback .... 1565 kg (3450 lb)
- Estate .... 1585 kg (3495 lb)

These weights are for manual transmission models with basic trim. Models with higher levels of trim or automatic transmission may weigh more

Trailer weights (maximum)
Astra/Kadett (up to 1984):
- Unbraked trailer .... 450 kg (992 lb)
- Braked trailer, Hatchback .... 800 kg (1764 lb) (four-speed)/750 kg (1654 lb) (five speed)
- Braked trailer, Estate .... 700 kg (1544 lb) (four-speed)/650 kg (1433 lb) (five-speed)

Astra Kadett (1985 on):
- Unbraked trailer .... 450 kg (992 lb)
- Braked trailer, Hatchback .... 750 kg (1654 lb) (four-speed)/850 kg (1874 lb) (five-speed)
- Braked trailer, Estate .... 750 kg (1654 lb)
- Braked trailer, Van .... 750 kg (1654 lb)

Belmont:
- Unbraked trailer .... 450 kg (992 lb)
- Braked trailer .... 750 kg (1654 lb)

Cavalier/Ascona:
- Unbraked trailer .... 515 kg (1136 lb)
- Braked trailer, Saloon/Hatchback .... 750 kg (1654 lb)
- Braked trailer, Estate .... 700 kg (1544 lb)

Roof rack load
All models except Cavalier .... 100 kg (220 lb)
Cavalier .... 80 kg (176 lb)

Capacities
Engine oil, drain and refill, including filter:
- Up to 1984 .... 3.75 litre (6.6 pint) approx
- 1985 on* .... 5.00 litre (8.8 pint) approx

Cooling system, drain and refill .... 7.7 litre (13.6 pint) approx

Fuel tank:
- Astra/Kadett up to 1984 .... 42.0 litre (9.2 gallon) approx
- Astra/Kadett 1985 on:
  - Hatchback/Saloon .... 52.0 litre (11.4 gallon) approx
  - Estate/Van .... 50.0 litre (11.0 gallon) approx
- Belmont .... 52.0 litre (11.4 gallon) approx
- Cavalier/Ascona .... 61.0 litre (13.4 gallon) approx

Cavalier/Ascona engines with increased sump capacity are identified by having a green dipstick

Buying spare parts

Only GM spare parts should be used if the vehicle (or engine) is still under warranty. The use of other makes of part may invalidate the warranty if a claim has to be made. In any case, only buy parts of reputable make. Pattern parts, often of unknown origin, may not conform to the maker's standards either dimensionally or in material quality.

Large items or sub-assemblies – eg cylinder heads, starter motors, injection pumps – may be available on an 'exchange' basis. Consult a GM dealer for availablity and conditions. Dismantled or badly damaged units may not be accepted in exchange.

When buying engine parts, be prepared to quote the engine number. This is stamped onto the block just behind the top of the dipstick tube (photo). From 1986, the engine will normally carry a prefix 16DA to the engine number. On some early 1986 models, a prefix 16D was used with the suffix DA. These engines are of the later reduced exhaust emissions type. For purchases of a more general nature, the vehicle identification number (VIN) may be needed; this is stamped on a plate secured to the bonnet lock platform (photo).

Engine number is stamped on block (arrowed) behind dipstick tube

Vehicle identification plate

General repair procedures

Whenever servicing, repair or overhaul work is carried out on the car or its components, it is necessary to observe the following procedures and instructions. This will assist in carrying out the operation efficiently and to a professional standard of workmanship.

Joint mating faces and gaskets

Where a gasket is used between the mating faces of two components, ensure that it is renewed on reassembly, and fit it dry unless otherwise stated in the repair procedure. Make sure that the mating faces are clean and dry with all traces of old gasket removed. When cleaning a joint face, use a tool which is not likely to score or damage the face, and remove any burrs or nicks with an oilstone or fine file.

Make sure that tapped holes are cleaned with a pipe cleaner, and keep them free of jointing compound if this is being used unless specifically instructed otherwise.

Ensure that all orifices, channels or pipes are clear and blow through them, preferably using compressed air.

Oil seals

Whenever an oil seal is removed from its working location, either individually or as part of an assembly, it should be renewed.

The very fine sealing lip of the seal is easily damaged and will not seal if the surface it contacts is not completely clean and free from scratches, nicks or grooves. If the original sealing surface of the component cannot be restored, the component should be renewed.

Protect the lips of the seal from any surface which may damage them in the course of fitting. Use tape or a conical sleeve where possible. Lubricate the seal lips with oil before fitting and, on dual lipped seals, fill the space between the lips with grease.

Unless otherwise stated, oil seals must be fitted with their sealing lips toward the lubricant to be sealed.

Use a tubular drift or block of wood of the appropriate size to install the seal and, if the seal housing is shouldered, drive the seal down to the shoulder. If the seal housing is unshouldered, the seal should be fitted with its face flush with the housing top face.

Screw threads and fastenings

Always ensure that a blind tapped hole is completely free from oil, grease, water or other fluid before installing the bolt or stud. Failure to do this could cause the housing to crack due to the hydraulic action of the bolt or stud as it is screwed in.

When tightening a castellated nut to accept a split pin, tighten the nut to the specified torque, where applicable, and then tighten further to the next split pin hole. Never slacken the nut to align a split pin hole unless stated in the repair procedure.

When checking or retightening a nut or bolt to a specified torque setting, slacken the nut or bolt by a quarter of a turn, and then retighten to the specified setting.

Locknuts, locktabs and washers

Any fastening which will rotate against a component or housing in the course of tightening should always have a washer between it and the relevant component or housing.

Spring or split washers should always be renewed when they are used to lock a critical component such as a big-end bearing retaining nut or bolt.

Locktabs which are folded over to retain a nut or bolt should always be renewed.

Self-locking nuts can be reused in non-critical areas, providing resistance can be felt when the locking portion passes over the bolt or stud thread.

Split pins must always be replaced with new ones of the correct size for the hole.

Special tools

Some repair procedures in this manual entail the use of special tools such as a press, two or three-legged pullers, spring compressors etc. Wherever possible, suitable readily available alternatives to the manufacturer's special tools are described, and are shown in use. In some instances, where no alternative is possible, it has been necessary to resort to the use of a manufacturer's tool and this has been done for reasons of safety as well as the efficient completion of the repair operation. Unless you are highly skilled and have a thorough understanding of the procedure described, never attempt to bypass the use of any special tool when the procedure described specifies its use. Not only is there a very great risk of personal injury, but expensive damage could be caused to the components involved.

Tools and working facilities

Introduction

A selection of good tools is a fundamental requirement for anyone contemplating the maintenance and repair of a motor vehicle. For the owner who does not possess any, their purchase will prove a considerable expense, offsetting some of the savings made by doing-it-yourself. However, provided that the tools purchased meet the relevant national safety standards and are of good quality, they will last for many years and prove an extremely worthwhile investment.

To help the average owner to decide which tools are needed to carry out the various tasks detailed in this manual, we have compiled three lists of tools under the following headings: *Maintenance and minor repair*, *Repair and overhaul*, and *Special*. The newcomer to practical mechanics should start off with the *Maintenance and minor repair* tool kit and confine himself to the simpler jobs around the vehicle. Then, as his confidence and experience grow, he can undertake more difficult tasks, buying extra tools as, and when, they are needed. In this way, a *Maintenance and minor repair* tool kit can be built-up into a *Repair and overhaul* tool kit over a considerable period of time without any major cash outlays. The experienced do-it-yourselfer will have a tool kit good enough for most repair and overhaul procedures and will add tools from the *Special* category when he feels the expense is justified by the amount of use to which these tools will be put.

Maintenance and minor repair tool kit

The tools given in this list should be considered as a minimum requirement if routine maintenance, servicing and minor repair operations are to be undertaken. We recommend the purchase of combination spanners (ring one end, open-ended the other); although more expensive than open-ended ones, they do give the advantages of both types of spanner.

Combination spanners - 10, 11, 12, 13, 14 & 17 mm
Adjustable spanner - 9 inch
Brake bleed nipple spanner
Screwdriver - 4 in long x $1/4$ in dia (flat blade)
Screwdriver - 4 in long x $1/4$ in dia (cross blade)
Combination pliers - 6 inch
Hacksaw (junior)
Tyre pump
Tyre pressure gauge
Oil can
Fine emery cloth (1 sheet)
Wire brush (small)
Funnel (medium size)
Chain or strap wrench

Repair and overhaul tool kit

These tools are virtually essential for anyone undertaking any major repairs to a motor vehicle, and are additional to those given in the *Maintenance and minor repair* list. Included in this list is a comprehensive set of sockets. Although these are expensive they will be found invaluable as they are so versatile - particularly if various drives are included in the set. We recommend the $1/2$ in square-drive type, as this can be used with most proprietary torque wrenches. If you cannot afford a socket set, even bought piecemeal, then inexpensive tubular box spanners are a useful alternative.

The tools in this list will occasionally need to be supplemented by tools from the *Special* list.

Sockets (or box spanners) to cover range in previous list, plus 27 mm/$1^1/16$ in for injectors
Reversible ratchet drive (for use with sockets)
Extension piece, 10 inch (for use with sockets)
Universal joint (for use with sockets)
Torque wrench (for use with sockets)
'Mole' wrench - 8 inch
Ball pein hammer
Soft-faced hammer, plastic or rubber
Screwdriver - 6 in long x $5/16$ in dia (flat blade)
Screwdriver - 2 in long x $5/16$ in square (flat blade)
Screwdriver - $1^1/2$ in long x $1/4$ in dia (cross blade)
Screwdriver - 3 in long x $1/8$ in dia (electricians)
Pliers - electricians side cutters
Pliers - needle nosed
Pliers - circlip (internal and external)
Cold chisel - $1/2$ inch
Scriber
Scraper
Centre punch
Pin punch
Hacksaw
Valve grinding tool
Steel rule/straight-edge
Allen keys
Dial test indicator and stand (photo)
Selection of files
Wire brush (large)
Axle-stands
Jack (strong trolley or hydraulic type)
Light with extension lead

Special tools

The tools in this list are those which are not used regularly, are expensive to buy, or which need to be used in accordance with their manufacturers' instructions. Unless relatively difficult mechanical jobs

A dial test indicator and stand will be needed for some operations on the engine and injection system

Tools and working facilities

are undertaken frequently, it will not be economic to buy many of these tools. Where this is the case, you could consider clubbing together with friends (or joining a motorists' club) to make a joint purchase, or borrowing the tools against a deposit from a local garage or tool hire specialist.

The following list contains only those tools and instruments freely available to the public, and not those special tools produced by the vehicle manufacturer specifically for its dealer network. You will find occasional references to these manufacturers' special tools in the text of this manual. Generally, an alternative method of doing the job without the vehicle manufacturers' special tool is given. However, sometimes, there is no alternative to using them. Where this is the case and the relevant tool cannot be bought or borrowed, you will have to entrust the work to a franchised garage.

Valve spring compressor
Piston ring compressor
Balljoint separator
Universal hub/bearing puller
Impact screwdriver
Micrometer and/or vernier gauge
Universal electrical multi-meter
Cylinder compression gauge (suitable for Diesel)
Lifting tackle
Trolley jack
Camshaft drivebelt tension gauge

Buying tools

For practically all tools, a tool factor is the best source since he will have a very comprehensive range compared with the average garage or accessory shop. Having said that, accessory shops often offer excellent quality tools at discount prices, so it pays to shop around.

There are plenty of good tools around at reasonable prices, but always aim to purchase items which meet the relevant national safety standards. If in doubt, ask the proprietor or manager of the shop for advice before making a purchase.

Care and maintenance of tools

Having purchased a reasonable tool kit, it is necessary to keep the tools in a clean serviceable condition. After use, always wipe off any dirt, grease and metal particles using a clean, dry cloth, before putting the tools away. Never leave them lying around after they have been used. A simple tool rack on the garage or workshop wall, for items such as screwdrivers and pliers is a good idea. Store all normal wrenches and sockets in a metal box. Any measuring instruments, gauges, meters, etc, must be carefully stored where they cannot be damaged or become rusty.

Take a little care when tools are used. Hammer heads inevitably become marked and screwdrivers lose the keen edge on their blades from time to time. A little timely attention with emery cloth or a file will soon restore items like this to a good serviceable finish.

Working facilities

Not to be forgotten when discussing tools, is the workshop itself. If anything more than routine maintenance is to be carried out, some form of suitable working area becomes essential.

It is appreciated that many an owner mechanic is forced by circumstances to remove an engine or similar item, without the benefit of a garage or workshop. Having done this, any repairs should always be done under the cover of a roof.

Wherever possible, any dismantling should be done on a clean, flat workbench or table at a suitable working height.

Any workbench needs a vice: one with a jaw opening of 4 in (100 mm) is suitable for most jobs. As mentioned previously, some clean dry storage space is also required for tools, as well as for lubricants, cleaning fluids, touch-up paints and so on, which become necessary.

Another item which may be required, and which has a much more general usage, is an electric drill with a chuck capacity of at least 5/16 in (8 mm). This, together with a good range of twist drills, is virtually essential for fitting accessories such as mirrors and reversing lights.

Last, but not least, always keep a supply of old newspapers and clean, lint-free rags available, and try to keep any working area as clean as possible.

Spanner jaw gap comparison table

Jaw gap (in)	Spanner size
0.250	1/4 in AF
0.276	7 mm
0.313	5/16 in AF
0.315	8 mm
0.344	11/32 in AF; 1/8 in Whitworth
0.354	9 mm
0.375	3/8 in AF
0.394	10 mm
0.433	11 mm
0.438	7/16 in AF
0.445	3/16 in Whitworth; 1/4 in BSF
0.472	12 mm
0.500	1/2 in AF
0.512	13 mm
0.525	1/4 in Whitworth; 5/16 in BSF
0.551	14 mm
0.563	9/16 in AF
0.591	15 mm
0.600	5/16 in Whitworth; 3/8 in BSF
0.625	5/8 in AF
0.630	16 mm
0.669	17 mm
0.686	11/16 in AF
0.709	18 mm
0.710	3/8 in Whitworth; 7/16 in BSF
0.748	19 mm
0.750	3/4 in AF
0.813	13/16 in AF
0.820	7/16 in Whitworth; 1/2 in BSF
0.866	22 mm
0.875	7/8 in AF
0.920	1/2 in Whitworth; 9/16 in BSF
0.938	15/16 in AF
0.945	24 mm
1.000	1 in AF
1.010	9/16 in Whitworth; 5/8 in BSF
1.024	26 mm
1.063	1 1/16 in AF; 27 mm
1.100	5/8 in Whitworth; 11/16 in BSF
1.125	1 1/8 in AF
1.181	30 mm
1.200	11/16 in Whitworth; 3/4 in BSF
1.250	1 1/4 in AF
1.260	32 mm
1.300	3/4 in Whitworth; 7/8 in BSF
1.313	1 5/16 in AF
1.390	13/16 in Whitworth; 15/16 in BSF
1.417	36 mm
1.438	1 7/16 in AF
1.480	7/8 in Whitworth; 1 in BSF
1.500	1 1/2 in AF
1.575	40 mm; 15/16 in Whitworth
1.614	41 mm
1.625	1 5/8 in AF
1.670	1 in Whitworth; 1 1/8 in BSF
1.688	1 11/16 in AF
1.811	46 mm
1.813	1 13/16 in AF
1.860	1 1/8 in Whitworth; 1 1/4 in BSF
1.875	1 7/8 in AF
1.969	50 mm
2.000	2 in AF
2.050	1 1/4 in Whitworth; 1 3/8 in BSF
2.165	55 mm
2.362	60 mm

Safety first!

Professional motor mechanics are trained in safe working procedures. However enthusiastic you may be about getting on with the job in hand, do take the time to ensure that your safety is not put at risk. A moment's lack of attention can result in an accident, as can failure to observe certain elementary precautions.

There will always be new ways of having accidents, and the following points do not pretend to be a comprehensive list of all dangers; they are intended rather to make you aware of the risks and to encourage a safety-conscious approach to all work you carry out on your vehicle.

Essential DOs and DON'Ts

DON'T rely on a single jack when working underneath the vehicle. Always use reliable additional means of support, such as axle stands, securely placed under a part of the vehicle that you know will not give way.

DON'T attempt to loosen or tighten high-torque nuts (e.g. wheel hub nuts) while the vehicle is on a jack; it may be pulled off.

DON'T start the engine without first ascertaining that the transmission is in neutral (or 'Park' where applicable) and the parking brake applied.

DON'T suddenly remove the filler cap from a hot cooling system – cover it with a cloth and release the pressure gradually first, or you may get scalded by escaping coolant.

DON'T attempt to drain oil until you are sure it has cooled sufficiently to avoid scalding you.

DON'T grasp any part of the engine or exhaust without first ascertaining that it is sufficiently cool to avoid burning you.

DON'T allow brake fluid or antifreeze to contact vehicle paintwork.

DON'T syphon toxic liquids such as fuel, brake fluid or antifreeze by mouth, or allow them to remain on your skin.

DON'T inhale dust – it may be injurious to health (see *Asbestos* below).

DON'T allow any spilt oil or grease to remain on the floor – wipe it up straight away, before someone slips on it.

DON'T use ill-fitting spanners or other tools which may slip and cause injury.

DON'T attempt to lift a heavy component which may be beyond your capability – get assistance.

DON'T rush to finish a job, or take unverified short cuts.

DON'T allow children or animals in or around an unattended vehicle.

DO wear eye protection when using power tools such as drill, sander, bench grinder etc, and when working under the vehicle.

DO use a barrier cream on your hands prior to undertaking dirty jobs – it will protect your skin from infection as well as making the dirt easier to remove afterwards; but make sure your hands aren't left slippery. Note that long-term contact with used engine oil can be a health hazard.

DO keep loose clothing (cuffs, tie etc) and long hair well out of the way of moving mechanical parts.

DO remove rings, wristwatch etc, before working on the vehicle – especially the electrical system.

DO ensure that any lifting tackle used has a safe working load rating adequate for the job.

DO keep your work area tidy – it is only too easy to fall over articles left lying around.

DO get someone to check periodically that all is well, when working alone on the vehicle.

DO carry out work in a logical sequence and check that everything is correctly assembled and tightened afterwards.

DO remember that your vehicle's safety affects that of yourself and others. If in doubt on any point, get specialist advice.

IF, in spite of following these precautions, you are unfortunate enough to injure yourself, seek medical attention as soon as possible.

Asbestos

Certain friction, insulating, sealing, and other products – such as brake linings, brake bands, clutch linings, torque converters, gaskets, etc – contain asbestos. *Extreme care must be taken to avoid inhalation of dust from such products since it is hazardous to health.* If in doubt, assume that they *do* contain asbestos.

Fire

Remember at all times that fuel is highly flammable. Never smoke, or have any kind of naked flame around, when working on the vehicle. But the risk does not end there – a spark caused by an electrical short-circuit, by two metal surfaces contacting each other, by careless use of tools, or even by static electricity built up in your body under certain conditions, can ignite fuel vapour, which in a confined space is highly explosive.

Always disconnect the battery earth (ground) terminal before working on any part of the fuel or electrical system, and never risk spilling fuel on to a hot engine or exhaust.

It is recommended that a fire extinguisher of a type suitable for fuel and electrical fires is kept handy in the garage or workplace at all times. Never try to extinguish a fuel or electrical fire with water.

Note: *Any reference to a 'torch' appearing in this manual should always be taken to mean a hand-held battery-operated electric lamp or flashlight. It does NOT mean a welding/gas torch or blowlamp.*

Fumes

Certain fumes are highly toxic and can quickly cause unconsciousness and even death if inhaled to any extent. Fuel vapour comes into this category, as do the vapours from certain solvents such as trichloroethylene. Any draining or pouring of such volatile fluids should be done in a well ventilated area.

When using cleaning fluids and solvents, read the instructions carefully. Never use materials from unmarked containers – they may give off poisonous vapours.

Never run the engine of a motor vehicle in an enclosed space such as a garage. Exhaust fumes contain carbon monoxide which is extremely poisonous; if you need to run the engine, always do so in the open air or at least have the rear of the vehicle outside the workplace.

If you are fortunate enough to have the use of an inspection pit, never drain or pour fuel, and never run the engine, while the vehicle is standing over it; the fumes, being heavier than air, will concentrate in the pit with possibly lethal results.

The battery

Never cause a spark, or allow a naked light, near the vehicle's battery. It will normally be giving off a certain amount of hydrogen gas, which is highly explosive.

Always disconnect the battery earth (ground) terminal before working on the fuel or electrical systems.

If possible, loosen the filler plugs or cover when charging the battery from an external source. Do not charge at an excessive rate or the battery may burst.

Take care when topping up and when carrying the battery. The acid electrolyte, even when diluted, is very corrosive and should not be allowed to contact the eyes or skin.

If you ever need to prepare electrolyte yourself, always add the acid slowly to the water, and never the other way round. Protect against splashes by wearing rubber gloves and goggles.

When jump starting a car using a booster battery, for negative earth (ground) vehicles, connect the jump leads in the following sequence: First connect one jump lead between the positive (+) terminals of the two batteries. Then connect the other jump lead first to the negative (–) terminal of the booster battery, and then to a good earthing (ground) point on the vehicle to be started, at least 18 in (45 cm) from the battery if possible. Ensure that hands and jump leads are clear of any moving parts, and that the two vehicles do not touch. Disconnect the leads in the reverse order.

Mains electricity and electrical equipment

When using an electric power tool, inspection light etc, always ensure that the appliance is correctly connected to its plug and that, where necessary, it is properly earthed (grounded). Do not use such appliances in damp conditions and, again, beware of creating a spark or applying excessive heat in the vicinity of fuel or fuel vapour. Also ensure that the appliances meet the relevant national safety standards.

Diesel fuel

Diesel injection pumps supply fuel at very high pressure, and extreme care must be taken when working on the fuel injectors and fuel pipes. It is advisable to place an absorbent cloth around the union before slackening a fuel pipe and *never expose the hands or any part of the body to injector spray, as working pressure can cause the fuel to penetrate the skin with possibly fatal results.*

Routine maintenance

For modifications, and information applicable to later models, see Supplement at end of manual

The maintenance schedules below are basically those recommended by the manufacturer. The intervals specified for some items vary dramatically according to year, and the sceptic may wonder why, for instance, a fuel filter which must be renewed every 6000 miles on a 1983 vehicle will last for 18 000 miles on a 1984 vehicle. The extended oil change interval on later models is made possible by increasing the sump capacity, so should not be be applied to earlier models.

Some maintenance items should be performed more frequently on vehicles operating under adverse conditions. 'Adverse conditions' include extremes of climate, full-time towing, mainly taxi or short-haul work, and driving on unmade roads. Consult a GM dealer if in doubt on this point. Older or high-mileage vehicles may also need attention more frequently.

Where the annual mileage is low, the time interval should be used to determine when maintenance is due. This is because some systems and fluids deteriorate with time as well as with mileage.

Generally, only those tasks peculiar to Diesel-engined models are described in this manual. Reference should also be made to the appropriate manual for petrol-engined vehicles, where tasks common to both will be described.

Under-bonnet view of a GM Diesel-powered vehicle

1 Cooling system filler/pressure cap
2 Brake fluid reservoir cap
3 Windscreen wiper motor
4 Heater blower motor
5 Fuel hoses
6 Brake servo non-return valve
7 Fuel filter
8 Screen washer reservoir
9 Suspension turret
10 Air cleaner housing
11 Gearbox breather
12 Vacuum pump
13 Battery
14 Earth strap
15 Radiator fan
16 Engine oil dipstick
17 Engine oil filler
18 Fuel injection pump
19 Thermostat elbow
20 Engine breather
21 Cooling system vent hoses
22 Crankcase ventilation hose
23 Camshaft drivebelt cover
24 Coolant hose (to expansion tank)
25 Inlet manifold

Front three-quarter view of GM Diesel engine

1 Sump
2 Oil pressure regulator valve
3 Oil filter
4 Crankshaft pulley
5 Camshaft drivebelt cover
6 Engine mounting
7 Intake manifold
8 Breather hose
9 Thermostat elbow
10 Fuel inlet union
11 Fuel return union
12 Air cleaner intake
13 Oil filler cap
14 Oil dipstick
15 Vacuum pump
16 Fuel injector
17 Breather hose
18 Glow plug bus bar
19 Core plug
20 Flywheel
21 Fuel injection pump

Routine maintenance

Every 250 miles, weekly, or before a long journey – all models

Check engine oil level and top up if necessary (Chapter 1, Section 2)
Check coolant level and top up if necessary (Chapter 2, Section 2)
Check tyre pressures, including spare (Chapter 6, Section 2)
Top up screen washer reservoir(s)
Check operation of lights, wipers, washers etc

Schedule A – 1982/83 Model Years

Every 3000 miles

Renew engine oil and oil filter (Chapter 1, Section 4)
Drain fuel filter (Chapter 3, Section 2)

Every 6000 miles or 6 months, whichever comes first

Renew engine oil and oil filter
Check idle speed and adjust if necessary (Chapter 3, Section 5)
Renew fuel filter and air filter elements (Chapter 3, Sections 3 and 4)
Check accessory drivebelt(s) for condition and tension (Chapter 7, Section 3)
Check coolant level and antifreeze strength (Chapter 2, Section 2)
Check cooling system hoses for condition and security
Check battery electrolyte level (if applicable) (Chapter 7, Section 2)
Check power steering fluid level (if applicable)
Check transmission oil level
Check clutch pedal stroke
Lubricate hinges, locks etc
Check headlamp beam alignment
Inspect brake pads and renew if worn
Inspect rear brake shoes and renew if worn
Adjust rear brakes (when applicable)
Check handbrake adjustment, correct if necessary
Check operation of load-dependent brake proportioning or regulating valve (Estate/Vans, as applicable)
Inspect tyres thoroughly (Chapter 6, Section 2)
Check exhaust system for security and condition
Inspect shock absorbers and check function
Inspect steering rack bellows, driveshaft boots etc
Check steering and suspension balljoints for wear
Check front wheel alignment
Inspect brake pipes and hoses
Check tightness of wheel bolts

Every 18 000 miles or 2 years, whichever comes first

Check condition and tension of camshaft drivebelt (Chapter 1, Section 5)

Every 36 000 miles or 4 years, whichever comes first

Renew camshaft drivebelt (Chapter 1, Section 5)

Every 60 000 miles or 5 years, whichever comes first

Renew glow plugs (Chapter 3, Section 12)

Schedule B – 1984/85/86 Model Years

Every 3000 miles (1984)/4500 miles (1985/86) or six months, whichever comes first

Renew engine oil and oil filter (Chapter 1, Section 4)
Check battery electrolyte level (if applicable) (Chapter 7, Section 2)
Drain fuel filter (Chapter 3, Section 2)

Every 9000 miles or 12 months, whichever comes first

Renew engine oil and oil filter
Check idle speed and adjust if necessary (Chapter 3, Section 5)
Inspect fuel lines and fuel tank for leaks
Check accessory drivebelt(s) for condition and tension (Chapter 7, Section 3)
Lubricate throttle linkage
Check coolant level and antifreeze strength (Chapter 2, Section 2)
Check cooling system hoses for condition and security
Check clutch pedal stroke
Lubricate hinges, locks etc
Check headlamp beam alignment
Inspect brake pads and renew if worn
Check operation of load-dependent brake proportioning or regulating valve (Estates/Vans, as applicable)
Inspect tyres thoroughly (Chapter 6, Section 2)
Inspect brake pipes and hoses
Check exhaust system for security and condition
Inspect steering rack bellows, driveshaft boots etc
Check steering and suspension balljoints for wear
Check tightness of wheel bolts

Every 18 000 miles or 2 years, whichever comes first

In addition to the work previously specified
Renew fuel filter and air cleaner elements (Chapter 3, Sections 3 and 4)
Inspect rear brake shoes and renew if worn
Check power steering fluid level (when applicable)
Check transmission oil level
Check condition and tension of camshaft drivebelt (Chapter 1, Section 5)

Every 36 000 miles or 4 years, whichever comes first

Renew camhaft drivebelt (Chapter 1, Section 5)

Every 63 000 miles or 7 years, whichever comes first

Renew glow plugs (Chapter 3, Section 12)

Schedule C – 1987 Model Year onwards

Every 4500 miles or six months, whichever comes first

Renew engine oil and filter (Chapter 1, Section 4)
Drain fuel filter (Chapter 3, Section 2)

Every 9000 miles or 12 months, whichever comes first

As for Schedule B (above)

Every 18 000 miles or 2 years, whichever comes first

As for Schedule B, but without inspecting the camshaft drivebelt

Routine maintenance

Every 27 000 miles/3 years (adverse conditions) or 36 000 miles/4 years (normal conditions)

Check condition and tension of camshaft drivebelt (Chapter 1, Section 5)

Every 54 000 miles or 6 years, whichever comes first

Renew camshaft drivebelt (adverse conditions only) (Chapter 1, Section 5)

Every 63 000 miles or 7 years, whichever comes first

Renew glow plugs (Chapter 3, Section 12)
Renew camshaft drivebelt (normal operating conditions) (Chapter 1, Section 5)

Seasonal and other maintenance – All Models

Annually, regardless of mileage

Renew brake hydraulic fluid by bleeding

Every two years, regardless of mileage

Renew engine coolant if wished (Chapter 2, Sections 3, 5 and 11)

If suffering smoking, knocking or power loss

Check injection system – see Chapter 3, Section 18

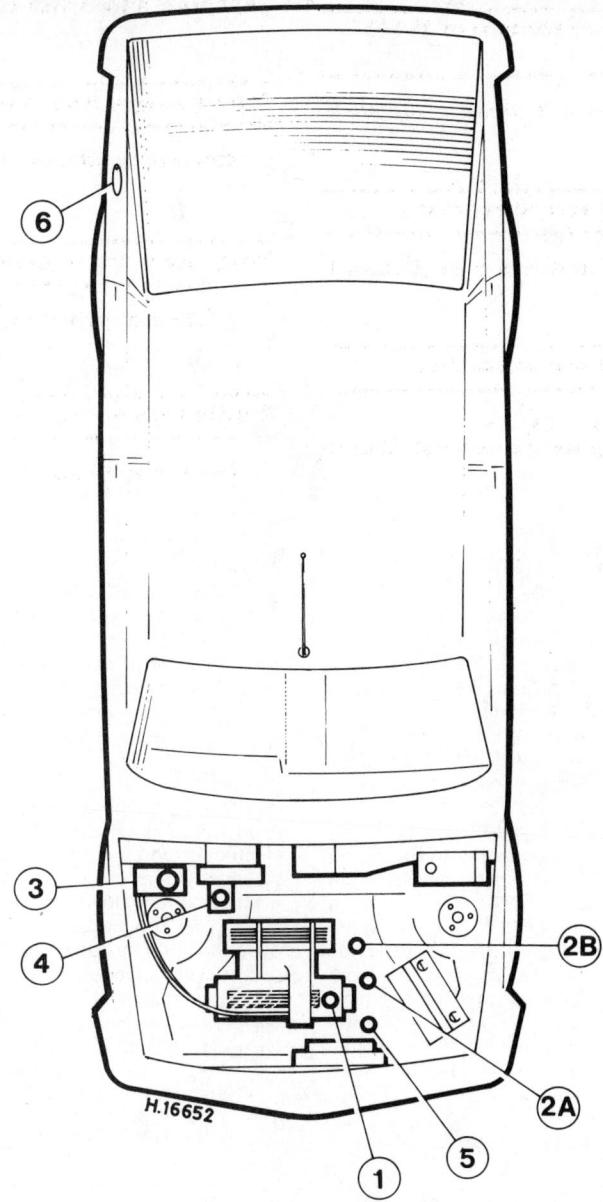

Recommended lubricants and fluids

Component or system	Lubricant type/specification	Duckhams recommendation
1 Engine	Multigrade engine oil, viscosity range SAE 10W/40 to 20W/50, to API SG/CD or better	Duckhams Diesel, QS, QXR, Hypergrade Plus or Hypergrade
2A Manual transmission	Gear oil, viscosity SAE 80 to GM 4753 M	Duckhams Hypoid 80W/90
2B Automatic transmission	Dexron II type ATF to GM 6137 M	Duckhams Uni-Matic
3 Cooling system	Ethylene glycol based antifreeze to GME 13368	Duckhams Universal Antifreeze and Summer Coolant
4 Brake hydraulic system	Hydraulic fluid to GME 05301 or GM 4653 M	Duckhams Universal Brake and Clutch Fluid
5 Power-assisted steering	Dexron II type ATF	Duckhams Uni-Matic
6 Fuel	Commercial Diesel fuel for road vehicles (DERV)	

Fault diagnosis

Introduction

The vehicle owner who does his or her own maintenance according to the recommended schedules should not have to use this section of the manual very often. Modern component reliability is such that, provided those items subject to wear or deterioration are inspected or renewed at the specified intervals, sudden failure is comparatively rare. Faults do not usually just happen as a result of sudden failure, but develop over a period of time. Major mechanical failures in particular are usually preceded by characteristic symptoms over hundreds or even thousands of miles. Those components which do occasionally fail without warning are often small and easily carried in the vehicle.

With any fault finding, the first step is to decide where to begin investigations. Sometimes this is obvious, but on other occasions a little detective work will be necessary. The owner who makes half a dozen haphazard adjustments or replacements may be successful in curing a fault (or its symptoms), but he will be none the wiser if the fault recurs and he may well have spent more time and money than was necessary. A calm and logical approach will be found to be more satisfactory in the long run. Always take into account any warning signs or abnormalities that may have been noticed in the period preceding the fault – power loss, high or low gauge readings, unusual noises or smells, etc – and remember that failure of components such as fuses may only be pointers to some underlying fault.

The pages which follow here are intended to help in cases of failure to start or breakdown on the road. There is also a Fault Diagnosis Section at the end of each Chapter which should be consulted if the preliminary checks prove unfruitful. Whatever the fault, certain basic principles apply. These are as follows:

Verify the fault. This is simply a matter of being sure that you know what the symptoms are before starting work. This is particularly important if you are investigating a fault for someone else who may not have described it very accurately.

Don't overlook the obvious. For example, if the vehicle won't start, is there fuel in the tank? (Don't take anyone else's word on this particular point, and don't trust the fuel gauge either!) If an electrical fault is indicated, look for loose or broken wires before digging out the test gear.

Cure the disease, not the symptom. Substituting a flat battery with a fully charged one will get you off the hard shoulder, but if the underlying cause is not attended to, the new battery will go the same way.

Don't take anything for granted. Particularly, don't forget that a 'new' component may itself be defective (especially if it's been rattling round in the boot for months), and don't leave components out of a fault diagnosis sequence just because they are new or recently fitted. When you do finally diagnose a difficult fault, you'll probably realise that all the evidence was there from the start.

Electrical faults

Electrical faults can be more puzzling than straightforward mechanical failures, but they are no less susceptible to logical analysis if the basic principles of operation are understood. Vehicle electrical wiring exists in extremely unfavourable conditions – heat, vibration and chemical attack – and the first things to look for are loose or corroded connections and broken or chafed wires, especially where the wires pass through holes in the bodywork or are subject to vibration.

All metal-bodied vehicles in current production have one pole of the battery 'earthed', ie connected to the vehicle bodywork, and in nearly all modern vehicles it is the negative (–) terminal. The various electrical components – motors, bulb holders etc – are also connected to earth, either by means of a lead or directly by their mountings. Electric current flows through the component and then back to the battery via the bodywork. If the component mounting is loose or corroded, or if a good path back to the battery is not available, the circuit will be incomplete and malfunction will result. The engine and/or gearbox are also earthed by means of flexible metal straps to the body or subframe; if these straps are loose or missing, starter motor, generator and other trouble may result.

Assuming the earth return to be satisfactory, electrical faults will be due either to component malfunction or to defects in the current supply. If supply wires are broken or cracked internally this results in an open-circuit, and the easiest way to check for this is to bypass the suspect wire temporarily with a length of wire having a crocodile clip or suitable connector at each end. Alternatively, a 12V test lamp can be used to verify the presence of supply voltage at various points along the wire and the break can be thus isolated.

If a bare portion of a live wire touches the bodywork or other earthed metal part, the electricity will take the low-resistance path thus formed back to the battery: this is known as a short-circuit. Hopefully a short-circuit will blow a fuse, but otherwise it may cause burning of the insulation (and possibly further short-circuits) or even a fire. This is why it is inadvisable to bypass persistently blowing fuses with silver foil or wire.

Spares and tool kit

Most vehicles are supplied only with sufficient tools for wheel changing; the *Maintenance and minor repair* tool kit detailed in *Tools and working facilities*, with the addition of a hammer, is probably sufficient for those repairs that most motorists would consider attempting at the roadside. In addition a few items which can be fitted without too much trouble in the event of a breakdown should be carried. Experience and available space will modify the list below, but the following may save having to call on professional assistance:

Drivebelt(s) – emergency type may suffice
Spare fuses
Set of principal light bulbs
Tin of radiator sealer and hose bandage

Fault diagnosis

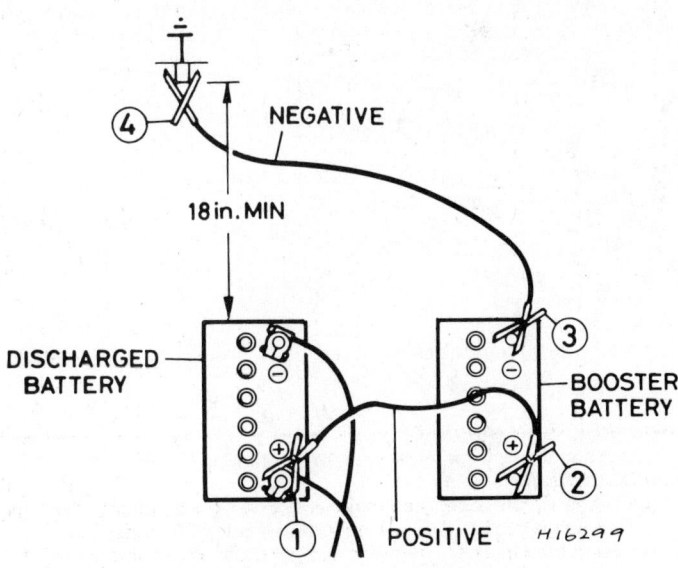

Jump start lead connections for negative earth vehicles – connect leads in order shown

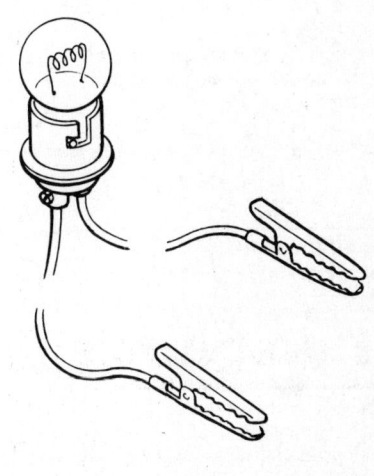

Simple test lamp is useful for tracing electrical faults

Exhaust bandage
Roll of insulating tape
Length of soft iron wire
Length of electrical flex
Torch or inspection lamp (can double as test lamp)
Battery jump leads
Tow-rope
Litre of engine oil
Sealed can of hydraulic fluid
Emergency windscreen
Worm drive clips

If spare fuel is carried, a can designed for the purpose should be used to minimise risks of leakage and collision damage. A first aid kit and a warning triangle, whilst not at present compulsory in the UK, are obviously sensible items to carry in addition to the above.

When touring abroad it may be advisable to carry additional spares which, even if you cannot fit them yourself, could save having to wait while parts are obtained. The items below may be worth considering:

Clutch and throttle cables
Cylinder head gasket
Alternator brushes
Tyre valve core
Fuel injector(s) and sealing washers

One of the motoring organisations will be able to advise on availability of fuel etc in foreign countries.

Starter motor turns engine slowly
  Partially discharged battery (recharge or use jump leads)
  Battery terminals loose or corroded
  Battery earth to body defective
  Engine earth strap loose
  Starter motor (or solenoid) wiring loose
  Starter motor internal fault (see Chapter 7)

Starter motor spins without turning engine
  Flywheel gear teeth damaged or worn
  Starter motor mounting bolts loose

Engine turns normally but fails to start
  No fuel in tank
  Incorrect use of cold start device
  Wax formed in fuel (in very cold weather)
  Exhaust system blocked
  Poor compression (see Chapter 1)
  Fuel system or preheater fault (see Chapter 3)
  Major mechanical failure

Engine fires but will not run
  Insufficient use of cold start device
  Preheater fault (see Chapter 3)
  Wax formed in fuel (in very cold weather)
  Other fuel system fault (see Chapter 3)

Engine will not start

Engine fails to turn when starter operated
  Flat battery (recharge or use jump leads)
  Battery terminals loose or corroded
  Battery earth to body defective
  Engine earth strap loose or broken
  Starter motor (or solenoid) wiring loose or broken
  Automatic transmission selector in wrong position, or inhibitor switch faulty
  Ignition/starter switch faulty
  Major mechanical failure (seizure)
  Starter or solenoid internal fault (see Chapter 7)

Engine cuts out and will not restart

Engine misfires before cutting out – fuel fault
  Fuel tank empty
  Fuel pump defective or filter blocked (check for delivery)
  Fuel tank filler vent blocked (suction will be evident on releasing cap)
  Other fuel system fault (see Chapter 3)

Engine cuts out – other causes
  Serious overheating
  Major mechanical failure (eg camshaft drive)

Engine overheats

    Coolant loss due to internal or external leakage (see Chapter 2)
    Thermostat defective
    Low oil level
    Brakes binding
    Radiator clogged externally or internally
    Electric cooling fan not operating correctly
    Engine waterways clogged

Note: *Do not add cold water to an overheated engine or damage may result*

Low engine oil pressure

Gauge reads low or warning light illuminated with engine running
    Oil level low or incorrect grade
    Defective gauge or sender unit
    Wire to sender unit earthed
    Engine overheating
    Oil filter clogged or bypass valve defective
    Oil pressure relief valve defective
    Oil pick-up strainer clogged
    Oil pump worn
    Worn main or big-end bearings

Note: *Low oil pressure in a high-mileage engine at tickover is not necessarily a cause for concern. Sudden pressure loss at speed is far more significant. In any event, check the gauge or warning light sender before condemning the engine.*

Engine noises

To inexperienced ears the Diesel engine sounds alarming even when there is nothing wrong with it. Try to have unusual noises expertly diagnosed before making renewals or repairs.

Whistling or wheezing noises
    Leaking vacuum hose
    Leaking manifold gasket
    Blowing head gasket

Tapping or rattling
    Worn valve gear
    Worn timing belt
    Broken piston ring (ticking noise)

Knocking or thumping
    Unintentional mechanical contact (eg fan blades)
    Worn drivebelt
    Peripheral component fault (generator, water pump etc)
    Fuel pump sucking air
    Fuel contaminated
    Fuel injector(s) leaking or sticking – see Chapter 3
    Worn big-end bearings (regular heavy knocking, perhaps less under load)
    Worn main bearings (rumbling and knocking, perhaps worsening under load)
    Piston slap (most noticeable when cold)

Chapter 1 Engine

For modifications, and information applicable to later models, see Supplement at end of manual

Contents

Ancillary components – removal	19
Camshaft – removal and refitting	8
Camshaft drivebelt – inspection, removal, refitting and tensioning	5
Cylinder head – dismantling and reassembly	22
Cylinder head – examination and overhaul	23
Cylinder head – removal and refitting	7
Cylinder head and pistons – decarbonising	24
Engine – complete dismantling	20
Engine – complete reassembly	28
Engine – initial start-up after overhaul	31
Engine – methods of removal	15
Engine and transmission – reconnection	29
Engine and transmission – refitting	30
Engine and transmission – removal	16
Engine and transmission – separation	17
Engine components – examination and renovation	26
Engine dismantling – general	18
Engine oil and oil filter – renewal	4
Engine reassembly – general	27
Engine/transmission mountings – renewal	14
Examination and renovation – general	25
Fault diagnosis – engine	32
Flywheel – removal and refitting	12
General description	1
Maintenance and inspection	2
Major operations possible with the engine in the vehicle	3
Oil pump – dismantling, overhaul and reassembly	21
Oil pump – removal and refitting	10
Oil seals – renewal	13
Pistons and connecting rods – removal and refitting	11
Sump – removal and refitting	9
Valve timing – checking and adjustment	6

Specifications

General

Engine type .... Four-cylinder, four stroke, overhead camshaft, indirect injection, compression ignition.

Maker's designation:
   1982 to 1985 .... 16D
   1986 on 16DA
Bore .... 80.00 mm (3.1496 in) nominal
Stroke .... 79.50 mm (3.1299 in)
Displacement .... 1598 cc (97.48 cu in) nominal
Firing order .... 1 – 3 – 4 – 2 (No 1 at pulley end)
Compression ratio .... 23 : 1
Performance:
   Maximum power .... 40 kW (54 bhp) @ 4600 rpm
   Maximum torque:
      16D engine .... 96 Nm (71 lbf ft) @ 2400 rpm
      16DA engine .... 95 Nm (69 lbf ft) @ 2400 rpm

Camshaft drivebelt

Tension (using gauge KM-510-A):
   New belt, warm 9.0
   New belt, cold 6.5
   Run-in belt, warm 8.0
   Run-in belt, cold 4.0

Chapter 1 Engine

Cylinder head

	Thickness (fitted)	Identification
Gasket thickness and identification:		
Piston projection up to 0.75 mm (0.030 in)	1.3 mm (0.051 in)	None
Piston projection 0.75 to 0.85 mm (0.030 to 0.034 in)	1.4 mm (0.055 in)	One notch
Piston projection above 0.85 mm (0.034 in)	1.5 mm (0.059 in)	Two notches

Valve clearance adjustment .... Automatic by hydraulic tappets

Valve seat width in head:
- Inlet .... 1.3 to 2.0 mm (0.051 to 0.079 in)
- Exhaust .... 1.3 to 2.6 mm (0.051 to 0.102 in)

Valve stem play in guide:
- Inlet .... 0.015 to 0.047 mm (0.0006 to 0.0019 in)
- Exhaust .... 0.030 to 0.062 mm (0.0012 to 0.0024 in)

Recess of valve head when fitted .... 0.25 to 0.75 mm (0.010 to 0.030 in)
Swirl chamber projection .... 0.00 to 0.04 mm (0.00 to 0.0016 in)
Sealing surface finish – peak-to-valley height .... 0.025 mm (0.001 in) max

Overall height of head:
- Maximum .... 106.10 mm (4.1772 in)
- Minimum .... 105.75 mm (4.1634 in)

Deviation of sealing surface from true .... 0.15 mm (0.006 in) max

Valves

Length .... 123.25 mm (4.8524 in)

Head diameter:
- Inlet .... 36 mm (1.417 in)
- Exhaust .... 32 mm (1.260 in)

Stem diameter (standard):
- Inlet .... 7.970 to 7.985 mm (0.3138 to 0.3144 in)
- Exhaust .... 7.955 to 7.970 mm (0.3132 to 0.3138 in)

Valve guide bore (standard) .... 8.000 to 8.017 mm (0.3150 to 0.3156 in)

Valve stem oversizes:
- Marked 1 .... + 0.075 mm (0.0030 in)
- Marked 2 .... + 0.150 mm (0.0059 in)
- Marked A (inlet) or K (exhaust) .... + 0.250 mm (0.0098 in)

Valve sealing face angle .... 44°

Camshaft and bearings

Journal diameters (standard):
- No 1 .... 42.455 to 42.470 mm (1.6715 to 1.6720 in)
- No 2 .... 42.705 to 42.720 mm (1.6813 to 1.6819 in)
- No 3 .... 42.955 to 42.970 mm (1.6911 to 1.6917 in)
- No 4 .... 43.205 to 43.220 mm (1.7010 to 1.7016 in)
- No 5 .... 43.455 to 43.470 mm (1.7108 to 1.7114 in)

Bearing diameters (standard):
- No 1 .... 42.500 to 42.525 mm (1.6732 to 1.6742 in)
- No 2 .... 42.750 to 42.775 mm (1.6831 to 1.6841 in)
- No 3 .... 43.000 to 42.025 mm (1.6929 to 1.6939 in)
- No 4 .... 43.250 to 43.275 mm (1.7028 to 1.7037 in)
- No 5 .... 43.500 to 43.525 mm (1.7126 to 1.7136 in)

Camshaft and bearing undersize .... − 0.1 mm (0.0039 in)

Camshaft identification marks:
- Letter .... B
- Colour – standard .... None
- Colour – undersize .... Violet

Camshaft run-out .... 0.03 mm (0.0012 in) max
Camshaft endfloat .... 0.04 to 0.14 mm (0.0016 to 0.0055 in)
Cam lift (inlet and exhaust) .... 6.12 mm (0.2409 in)

Cylinder bores

	Diameter*	Identification
Production grade 1	79.95 mm (3.1476 in)	5
	79.96 mm (3.1480 in)	6
	79.97 mm (3.1484 in)	7
Production grade 2	79.98 mm (3.1488 in)	8
	79.99 mm (3.1492 in)	99
	80.00 mm (3.1496 in)	00
	80.01 mm (3.1500 in)	01
	80.02 mm (3.1504 in)	02
	80.03 mm (3.1508 in)	03
Production grade 3	80.04 mm (3.1512 in)	04
	80.05 mm (3.1516 in)	05
	80.06 mm (3.1520 in)	06
	80.07 mm (3.1524 in)	07
	80.08 mm (3.1528 in)	08
	80.09 mm (3.1531 in)	09
	80.10 mm (3.1535 in)	1
Oversize (0.5 mm)	80.47 mm (3.1681 in)	7 + 0.5
	80.48 mm (3.1685 in)	8 + 0.5
	80.49 mm (3.1689 in)	9 + 0.5

Chapter 1 Engine

Oversize (1.0 mm)	80.50 mm (3.1693 in)	0 + 0.5
	80.97 mm (3.1878 in)	7 + 1.0
	80.98 mm (3.1882 in)	8 + 1.0
	80.99 mm (3.1886 in)	9 + 1.0
	81.00 mm (3.1890 in)	0 + 1.0

* Tolerance ± 0.005 mm (0.0002 in)
Bore out-of-round and taper .... 0.013 mm (0.0005 in) max

Pistons
Make .... Mahle or Alcan
Marking (maker's identity):
    Mahle .... ⊓
    Alcan .... △
Diameter:
    Mahle .... 0.030 mm (0.0012 in) less than bore diameter
    Alcan .... 0.020 mm (0.0008 in) less than bore diameter
Grade marking .... As for cylinder bore
Projection at TDC (used to determine head gasket thickness) .... 0.65 to 0.95 mm (0.0256 to 0.0374 in)
Clearance in bore:
    Mahle .... 0.020 to 0.040 mm (0.0008 to 0.0016 in)
    Alcan .... 0.015 to 0.035 mm (0.0006 to 0.0014 in)

Piston rings
Thickness:
    Compression rings .... 1.978 to 1.990 mm (0.0779 to 0.0784 in)
    Oil scraper ring .... 2.975 to 3.010 mm (0.1171 to 0.1185 in)
End gap (fitted) .... 0.2 to 0.4 mm (0.008 to 0.016 in)
Vertical clearance in groove .... Not stated
Gap offset .... 180°

Gudgeon pins
Length .... 64.7 to 65.0 mm (2.547 to 2.559 in)
Diameter .... 25.995 to 26.000 mm (1.0234 to 1.0236 in)
Clearance in piston .... 0.007 to 0.011 mm (0.0003 to 0.0004 in)
Clearance in connecting rod .... 0.014 to 0.025 mm (0.0006 to 0.0010 in)

Connecting rods
Weight variation (in same engine) .... 4 g (0.14 oz) max
Identification of weight class:
    785 g nominal Black/1
    789 g nominal Blue/2
    793 g nominal Green/3
    797 g nominal Yellow/4
    807 g nominal White/5
    805 g nominal Grey/6
Endfloat on crankshaft .... 0.07 to 0.24 mm (0.0028 to 0.0095 in)

Crankshaft and bearings
Main bearing journal diameter:
    Standard .... 57.982 to 57.995 mm (2.2828 to 2.2833 in)
    0.25 undersize (production and service) .... 57.732 to 57.745 mm (2.2729 to 2.2734 in)
    0.50 undersize (service only) .... 57.482 to 57.495 mm (2.2631 to 2.2636 in)
Big-end bearing journal diameter:
    Standard .... 48.971 to 48.987 mm (1.9280 to 1.9286 in)
    0.25 undersize .... 48.721 to 48.737 mm (1.9182 to 1.9188 in)
    0.50 undersize .... 48.471 to 48.487 mm (1.9083 to 1.9089 in)

Bearing shell identification – standard:

	Colour	Marking
Main, except centre, top	Brown	413 N or 403 N
Main, centre, top	Brown	400 N or 410 N
Main, except centre, bottom	Green	414 N or 404 N
Main, centre, bottom	Green	401 N or 411 N
Big-end	None	419 N

Bearing shells identification – 0.25 undersize:

	Colour	Marking
Main, except centre, top	Brown-blue	415 A or 405 A
Main, centre, top	Brown-blue	402 A or 412 A
Main, except centre, bottom	Green-blue	416 A or 406 A
Main, centre, bottom	Green-blue	403 A or 413 A
Big-end	Blue	420 A

Chapter 1 Engine

Bearing shell identification – 0.50 undersize:
    Main, except centre, top .... Brown-white   236 B or 407
    Main, centre, top .... Brown-white   238 B or 414 B
    Main, except centre, bottom .... Green-white   237 B or 408
    Main, centre, bottom .... Green-white   239 B or 415 B
    Big-end .... White   421 B
Crankshaft bearing out-of-round and taper .... 0.004 mm (0.00016 in) max
Crankshaft run-out (measured at centre main bearing journal) ... 0.03 mm (0.0012 in) max
Crankshaft endfloat .... 0.07 to 0.30 mm (0.0028 to 0.0118 in)
Main bearing running clearance .... 0.015 to 0.040 mm (0.0006 to 0.0016 in)
Big-end bearing running clearance .... 0.019 to 0.063 mm (0.0008 to 0.0025 in)

Flywheel
Run-out .... 0.5 mm (0.020 in) max
Refacing limit (depth of material removed from clutch wear face) ...... 0.3 mm (0.012 in) max

Lubrication system
System type .... Wet sump, pressure feed, full flow filter
Lubricant type/specification.. Multigrade engine oil, viscosity range SAE 10W/40 to 20W/50, to API SG/CD or better (Duckhams Diesel, QS, QXR, Hypergrade Plus or Hypergrade)
Lubricant capacity (drain and refill, including filter):
    Up to 1984 models .... 3.75 litre (6.6 pint) approx
    1985 model year and later* .... 5.00 litre (8.9 pint) approx
Oil filter type .... Champion G105
Oil pressure (at idle, engine warm) .... 1.5 bar (22 lbf/in^2)
Oil pump tolerances:
    Gear backlash .... 0.1 to 0.2 mm (0.004 to 0.008 in)
    Outer gear-to-housing clearance .... 0.03 to 0.10 mm (0.0012 to 0.0039 in)
    Gear recess below body .... 0.03 to 0.10 mm (0.0012 to 0.0039 in)
* Ascona/Cavalier engines identified by having a green dipstick

Torque wrench settings

	Nm	lbf ft
Crankshaft pulley-to-sprocket	20	15
Crankshaft pulley/sprocket centre bolt	155	114
Starter motor to block	45	33
Camshaft sprocket bolt:*		
Stage 1	75	55
Stage 2	Tighten a further 60°	
Oil pump housing to block	6	4
Flywheel to crankshaft:*		
Stage 1	60	44
Stage 2	Tighten a further 30°	
Main bearing caps	80	59
Connecting rod caps	50	37
Oil pump pressure regulator valve	30	22
Sump bolts	5	4
Cylinder head bolts:*		
Stage 1	25	18
Stage 2	Tighten a further 90°	
Stage 3	Tighten a further 90°	
Stage 4	Tighten a futher 45°	
Stage 5 (after warm-up)	Tighten a futher 30°	
Stage 6 (after 600 miles/1000 km)	Tighten a further 45°	
Vacuum pump to camshaft housing	28	21
Oil pressure switch	30	22
Crankcase ventilation oil separator to block	15	11
Engine mounting bracket (RH) to block	50	37
Other engine mountings	40	30
Oil drain plug	45	33

* Bolts tightened by the angular method should be renewed every time

1 General description

The engine is of four-cylinder overhead camshaft type. It bears many similarities to the family of petrol engines from which it is derived. The cylinder head is of light alloy construction and the block is of iron. The crankshaft runs in five main bearings; both main and big-end journals are hardened to withstand the greater loads imposed by compression ignition.

The camshaft is belt-driven from the crankshaft sprocket. The same belt drives the water pump and the fuel injection pump. Belt tension is

Chapter 1 Engine

adjusted by moving the water pump, which has its sprocket eccentrically mounted.

Lubrication is by means of a pump driven directly from the nose of the crankshaft. Oil is drawn from the sump and pumped through a full flow filter before entering the main oil galleries in the block and the crankshaft. Two pressure regulating valves, one in the pump and one in the cylinder head, open when the oil pressure exceeds a certain value. A bypass valve in the oil filter housing ensures a continued supply of oil, albeit unfiltered, should the filter element become clogged.

Valve clearance adjustment is automatic by means of the hydraulic lifters which support the cam followers. Both inlet and exhaust valve springs sit on rotators.

Crankcase ventilation is achieved by two hoses. One connects the lower crankcase to the camshaft carrier and runs between the carrier and the dipstick tube; the other runs from a filter in the camshaft cover to the inlet manifold side of the air cleaner.

Maintenance tasks are mostly straightforward, but operations which involve disturbing or checking the valve timing should not be undertaken unless the necessary equipment is available. Full details will be found in the appropriate Sections.

Apart from minor component differences, the information in respect of automatic transmission and power-assisted steering as installed in petrol-engined vehicles also applies to diesel-engined versions.

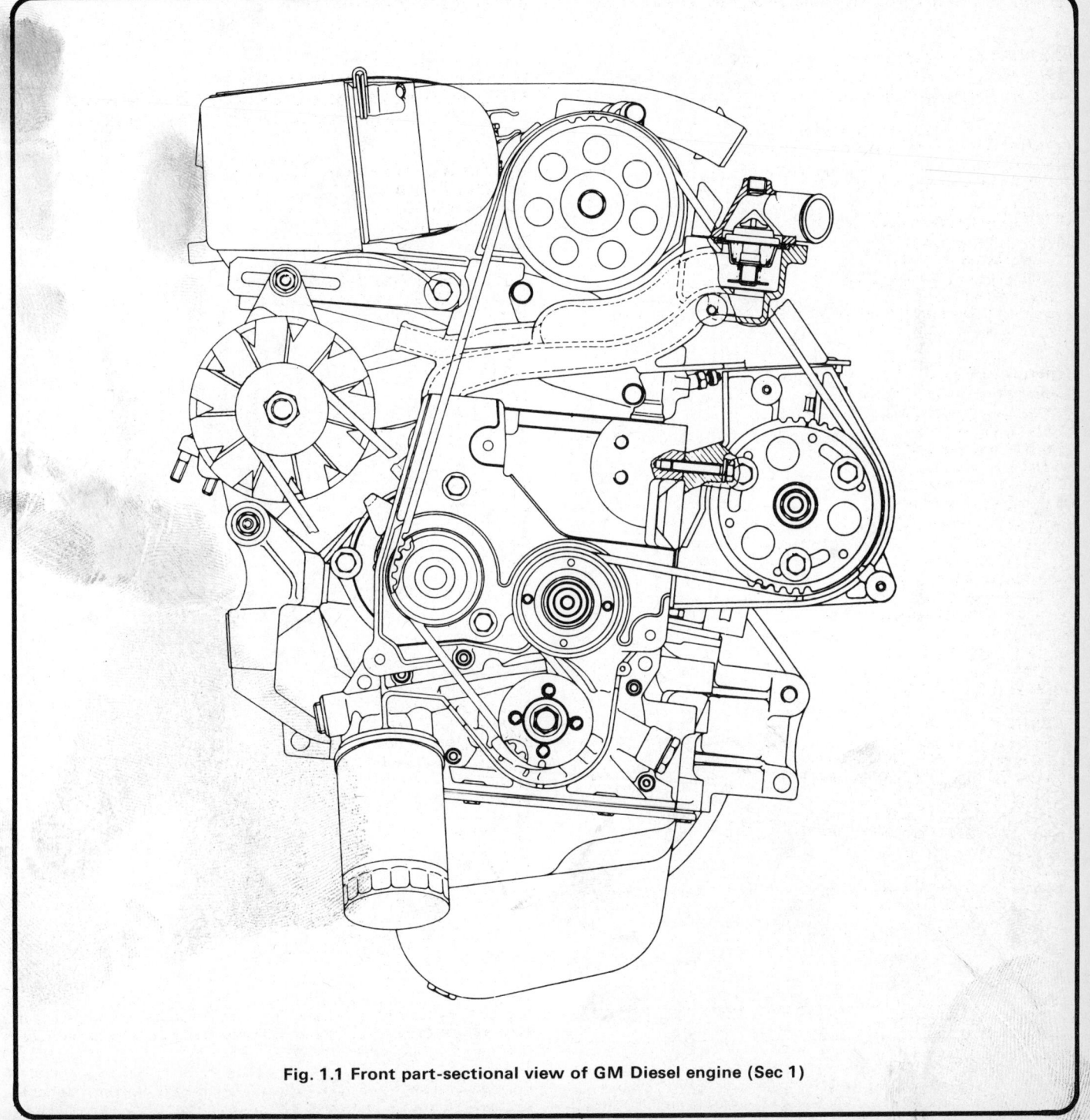

Fig. 1.1 Front part-sectional view of GM Diesel engine (Sec 1)

Chapter 1 Engine

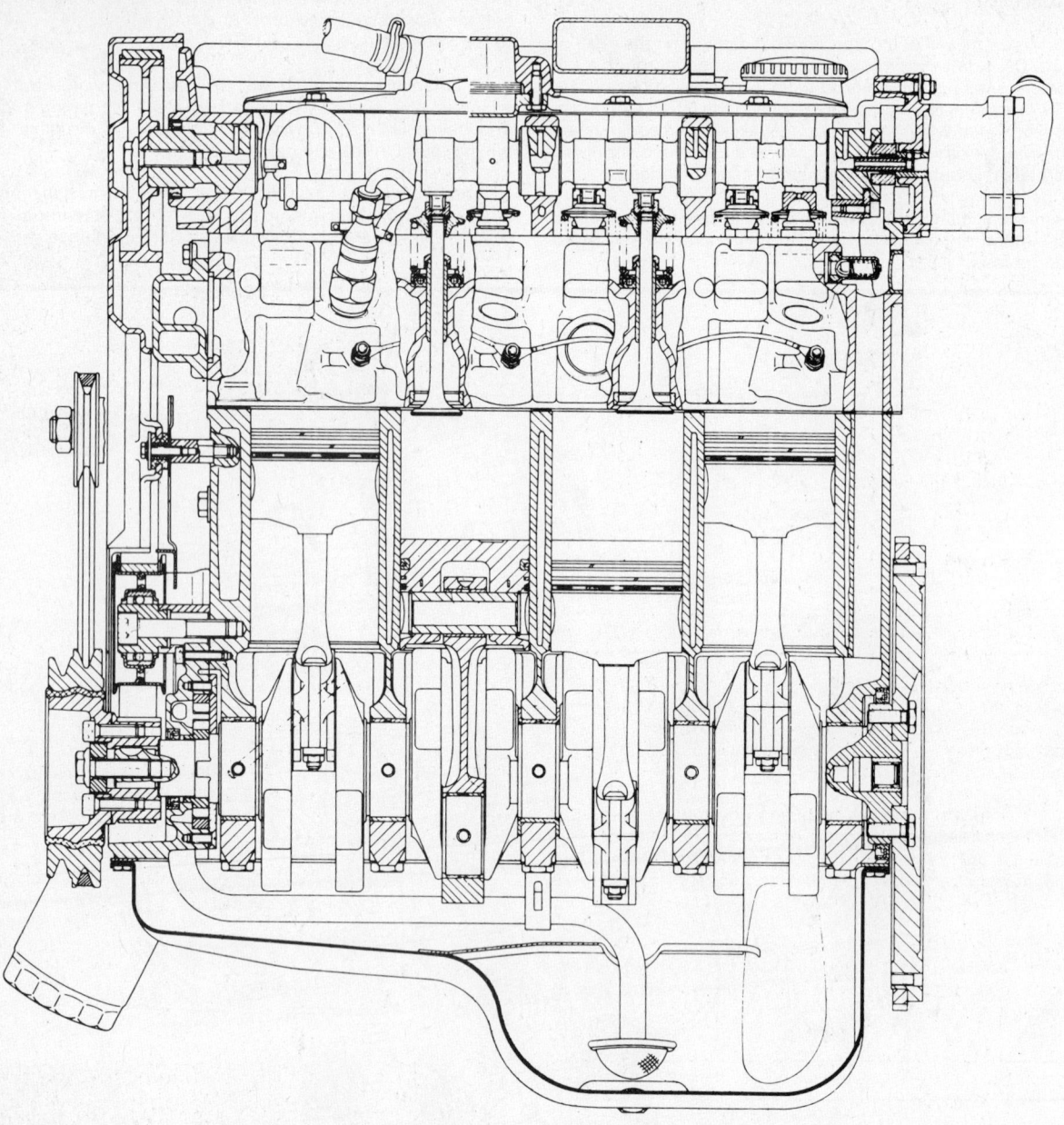

Fig. 1.2 Longitudinal section of GM Diesel engine (Sec 1)

2 Maintenance and inspection

1 The engine oil level should be checked weekly, before a long journey, or every 250 miles – more frequently if experience shows this to be necessary. The vehicle should be parked on level ground and the engine should have been stopped for a couple of minutes before checking the level. Open the bonnet and withdraw the dipstick. Wipe the dipstick clean, re-insert it fully, withdraw it again and read the oil level (photo). The level should be between the two marks on the dipstick; if it is approaching, or below, the low mark, add oil through the filler cap in the camshaft cover (photo). Do not overfill: the quantity needed to raise the level from the low to the high mark is approximately 1 litre (1.8 pints). Refit the dipstick and filler cap on completion, and mop up any oil spilt.

2 All engines use some oil; the makers state that consumption of up to 1.5 litres per 1000 km (4.25 pints per 1000 miles) is acceptable. Actual consumption will depend on driving conditions and styles. Check for leaks if frequent topping-up is required. If oil is not being lost through leaks, it is passing the valve stems and/or piston rings and being burnt.

3 At the specified intervals, or more frequently under adverse conditions (see Routine Maintenance), the engine oil and filter must be renewed as described in Section 4.

4 After an oil change, inspect the engine thoroughly for leaks of oil, coolant, fuel or products of combustion. Pay particular attention to joint faces and shaft sealing areas. Rectify leaks without delay.

5 Check the camshaft drivebelt tension at the specified intervals. See Section 5.

6 Renew the camshaft drivebelt as a precautionary measure at the specified intervals. Should the belt break in service, extensive damage may be caused. Refer to Section 5.

Chapter 1 Engine

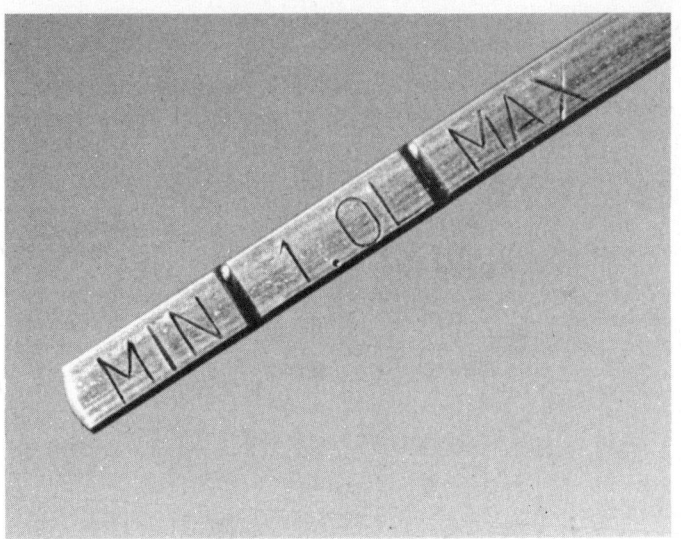

2.1A Engine oil dipstick markings

2.1B Topping up the engine oil

3 Major operations possible with the engine in the vehicle

1 Since the cylinder head, sump and (when applicable) clutch can all be removed with the engine in place, most dismantling can be carried out without removing the engine. However, access to some areas is difficult and if much work is to be done, it may be quicker in the long run to remove the engine.

2 Unless various special tools can be hired or borrowed, renewal of the crankshaft rear oil seal entails engine or transmission removal. This is unavoidable on automatic transmission models, since the torque converter cannot be removed with the engine and transmission installed.

3 In theory even the crankshaft and main bearings can be removed with the engine installed – again only on manual transmission models – but this is not recommended unless there is some compelling reason for not removing the engine.

4 Engine oil and oil filter – renewal

1 The engine oil should be drained when it is hot, so that it flows easily and its contaminants are in suspension. Take the vehicle for a brisk run if necessary to warm it up, then park on level ground and switch off the engine.

2 Place a drain pan of adequate capacity under the sump. Slacken and remove the sump drain plug and allow the oil to drain (photo). Clean the drain plug and its washer; renew the washer if it is in poor condition.

3 When the oil has finished draining, wipe clean around the drain hole, then refit the drain plug and washer. Tighten the drain plug.

4 The oil filter is at the rear of the engine at the pulley end; access is from below. Place the drain pan appropriately and unscrew the filter. If the filter will not unscrew by hand, use a strap or chain wrench or other proprietary tool. As a last resort, drive a screwdriver through the sides of the filter and use this as a lever to unscrew it. Be prepared for oil spillage.

5 Wipe clean on and around the filter seat and check that the sealing ring was not left behind. Smear the contact face of the sealing ring on the new filter with engine oil or grease.

6 Screw the new filter on, tightening it by hand only (photo). Typically the filter should be tightened by three-quarters of a turn beyond the point where the sealing ring makes contact with the housing; if more detailed tightening instructions are supplied with the new filter, follow them.

7 Fill the engine with the specified quantity and grade of oil. When the oil level appears correct on the dipstick, start the engine and run it for a few minutes. (The oil pressure light may not extinguish for a few seconds as the filter fills with oil). Check for leaks around the filter base and drain plug; tighten further if necessary.

8 Stop the engine, wait a few minutes and recheck the oil level. Top up if necessary to compensate for the oil absorbed by the new filter.

9 Put the old oil into a sealed container and dispose of it safely and in a non-polluting fashion.

4.2 Engine sump drain plug

4.6 Fitting a new oil filter

or contaminated; attend to the source of contamination too if necessary.

5 If it is only wished to check the belt tension, proceed to paragraph 19. To remove the belt, begin by removing the crankshaft pulley – it is secured to the sprocket by four Allen screws.

6 Disconnect the battery earth lead.

7 Remove the clutch/flywheel access cover from the bottom of the transmission bellhousing (photo).

8 Turn the crankshaft in the normal direction of rotation, using a spanner on the sprocket bolt, until the timing mark on the injection pump sprocket aligns with the reference mark on the pump bracket. In this position No 1 piston is at TDC on the firing stroke. See Chapter 3, photo 7.4.

9 Check that the TDC mark on the flywheel and the pointer on the clutch housing are aligned (photo).

10 If tool KM - 537 or equivalent is available, remove the vacuum pump and lock the camshaft in position by fitting the tool. If the tool is not available or cannot be fitted, make alignment marks between the camshaft sprocket and its backplate for use when refitting. See Section 6 for more details.

5 Camshaft drivebelt – inspection, removal, refitting and tensioning

1 The camshaft drivebelt (sometimes called the timing belt) also drives the water pump and the fuel injection pump. If it breaks in service, the pistons are likely to hit the valves.

2 Remove the alternator drivebelt – see Chapter 7, Section 3.

3 Remove the camshaft drivebelt covers. The large cover is secured by four screws – note the fuel pipe clip under one of them. The injection pump sprocket cover is secured by three screws (photo). (On 1987 models, the covers completely enclose the drivebelt; additional clamps and gaskets are fitted.)

4 Turn the engine using a spanner on the crankshaft pulley bolt – remove the right-hand front wheel for access if necessary. Inspect the belt for damage or contamination. Pay particular attention to the roots of the teeth, where cracking may occur. Renew the belt if it is damaged

11 Slacken the three bolts which secure the water pump to the block (photo). Using a large open-ended spanner on the flats of the pump, pivot it to release the tension on the belt.

12 Separate the right-hand front engine mounting by undoing the two bolts which are accessible from the top.

13 Mark the running direction of the belt if it is to be re-used. Also take care not to kink the belt, nor get oil, grease etc on it.

14 Slip the belt off the sprockets and jockey wheel. Remove the belt by feeding it through the engine mounting.

15 Commence refitting by threading the belt through the engine mounting. Refit and tighten the engine mounting bolts.

16 Place the belt over the sprockets and the jockey wheel. Make sure that No 1 piston is still at TDC, the injection pump sprocket mark is aligned and the camshaft position is still correct.

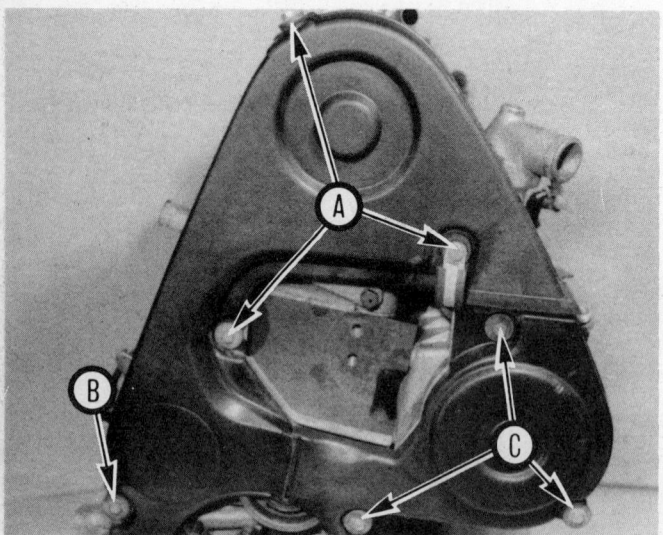

5.3 Camshaft drivebelt cover screws (engine removed)
A Large cover, short screws   C Pump sprocket cover screws
B Large cover, long screw

5.7 Removing the clutch/flywheel access cover

Chapter 1 Engine

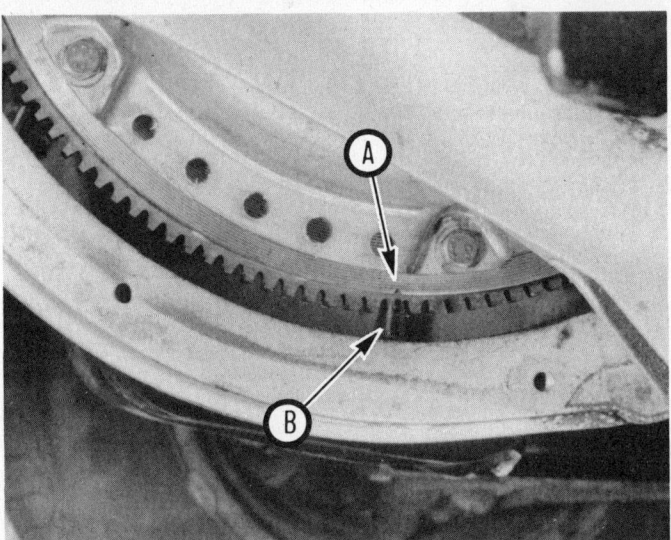

5.9 TDC mark on flywheel (A) and pointer on clutch housing (B)

5.11 Undoing a water pump bolt – other two arrowed (engine removed)

Fig. 1.3 Camshaft drivebelt correctly fitted (Sec 5)

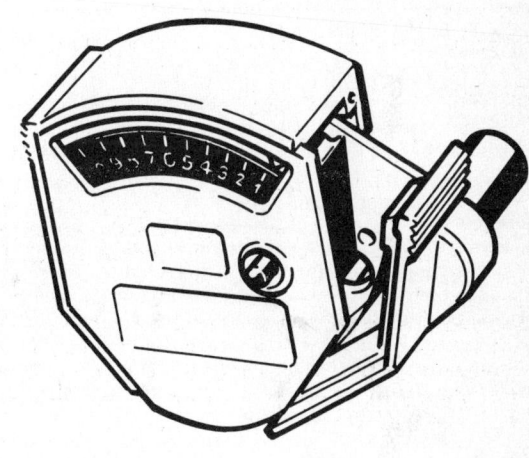

Fig. 1.4 Camshaft drivebelt tension gauge (Sec 5)

17 Move the water pump so as to put some tension on the drivebelt. Nip up the pump securing bolts, but do not tighten them fully yet.

18 Remove the camshaft locking tool, if used, and refit and secure the crankshaft pulley.

19 Belt tension can only be adjusted accurately using tension gauge KM-510-A or equivalent. If this gauge is not available, an approximation to the correct tension can be achieved by tensioning the belt so that it can just be twisted through 90° by thumb and forefinger in the middle of its longest run. A belt which is too tight will usually hum when running; a belt which is too slack will wear rapidly and may jump teeth. Use of a proper tension gauge is strongly recommended.

20 Settle the belt by rotating the crankshaft through half a turn in the normal direction of rotation. Fit the tension gauge to the slack side of the belt (the alternator side) and read the tension. Desired values are given in the Specifications.

21 If adjustment is necessary, slacken the water pump bolts and pivot the water pump to increase or decrease the tension. Nip up the water pump bolts.

22 Turn the crankshaft through one full turn, then recheck the tension. Keep adjusting the belt tension until a stable value is obtained.

23 Tighten the water pump bolts to the specified torque.

24 If the drivebelt has been re-tensioned or renewed, check the injection pump timing (Chapter 3, Section 7).

25 Refit the belt covers, clutch/flywheel cover and other disturbed components.

26 Refit the roadwheel, lower the vehicle and tighten the wheel bolts.

6 Valve timing – checking and adjustment

1 Valve timing on this engine is more complicated than on the petrol equivalent because there are no timing marks as such on the camshaft or sprocket, neither is the sprocket keyed or pegged to the camshaft. This makes the use of special tools unavoidable, although one can be home-made.

2   Note that the camshaft sprocket bolt should be renewed whenever it has been slackened.

3   If the valve timing has been lost completely, be careful when turning the crankshaft or camshaft in case piston/valve contact occurs.

4   Two methods of checking the valve timing are described. For either method, begin by checking the camshaft drivebelt tension as described in Section 5.

5   Bring the engine to TDC, No 1 firing – see Section 5, paragraphs 8 and 9.

6   Remove the air cleaner (Chapter 3, Section 4). Disconnect the breather hose and remove the camshaft cover (photo). If necessary, also remove the vacuum pump (Chapter 5, Section 3).

Using tool KM - 537 or equivalent
7   The maker's tool KM-537 consists of a plate which bolts onto the camshaft carrier in place of the vacuum pump; when the peg on the plate will enter the hole in the tail of the camshaft, the camshaft is correctly positioned for TDC, No 1 piston firing.

Making the tool
8   It is possible to make a substitute for tool KM-537, but the valve timing must be known to be correct first; thereafter the tool can be used to check the timing. As can be seen, the home-made tool is simply a metal bar with three holes drilled in it (photo). The accuracy of the tool depends on the precision with which the holes are drilled, and the snug fit of the screws or bolts in their holes. The crankshaft must be at TDC, No 1 firing, before marking up and constructing the tool.

9   Drill one of the end holes in the metal bar. Insert a short stud or dowel into the camshaft peg hole, marking the end with chalk or paint. The stud should be just long enough to touch the metal bar in its fitted position.

10 Secure the bar, using one of the vacuum pump screws, so that it is just free to move. Swing the bar past the stud or dowel so that an arc is marked on the bar. Remove the bar and drill a hole in the centre of the arc to accept a bolt or screw which will fit snugly into the camshaft peg hole. Fit this bolt or screw and clamp it with a couple of nuts.

11 Mark around the other vacuum pump screw hole with paint or chalk. Offer the tool to the camshaft and carrier so that the peg hole bolt and first fixing screw are snug and the position of the second fixing screw hole is marked. Remove the tool and drill the second fixing screw hole.

12 Offer up the tool again and make sure that it fits without strain or slack; repeat the construction exercise if necessary.

Using the tool
13 With the crankshaft and injection pump timing marks correctly positioned (paragraph 5), offer the tool to the camshaft carrier in place of the vacuum pump. If the tool can be located and secured so that the peg enters the hole in the camshaft, valve timing is correct. (In fact a tolerance of 1 mm/0.04 in either way at the flywheel TDC mark is allowed).

14 If the tool will not enter, remove it. Hold the camshaft using a spanner on the flats provided and slacken the camshaft sprocket bolt. Break the taper between the sprocket and camshaft if necessary by tapping the sprocket with a wooden or plastic mallet. Turn the camshaft until the alignment tool enters snugly. Remove the old camshaft sprocket bolt and insert a new bolt; nip the bolt up until the sprocket taper bites, then remove the alignment tool. Hold the camshaft again and tighten the sprocket bolt to the specified torque.

15 Back off the crankshaft a quarter turn, then regain TDC and check that the alignment tool still fits snugly.

Using a dial test indicator
16 Tool KM - 537 cannot be used on later models (mid-1984 on) because the holes into which it screws have been deleted. The

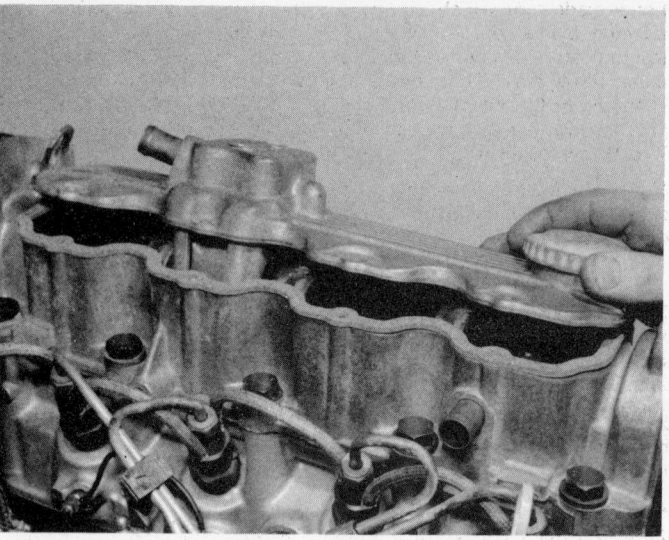
6.6 Removing the camshaft cover

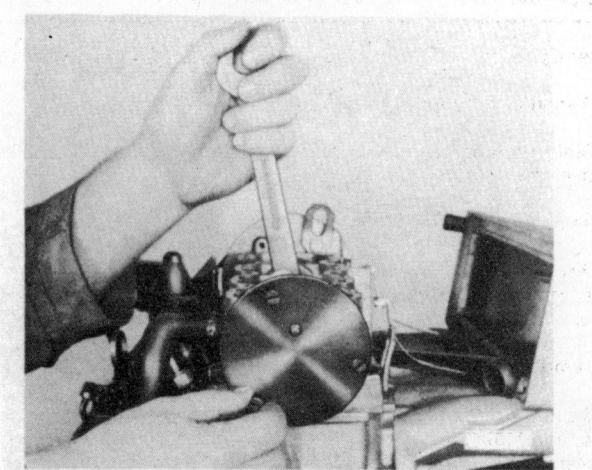
Fig. 1.5 Fitting the camshaft alignment tool KM-537 (Sec 6)

6.8 Home-made tool for locking the camshaft in a set position

home-made tool which screws into the vacuum pump holes is not affected. Instead of tool KM - 537, the makers specify the use of a dial test indicator (DTI or 'clock gauge') and suitable support. The support must allow the DTI to move across the camshaft carrier without changing height relative to the carrier top surface. The DTI 'foot' (the part which will rest on the cam lobe) should have a flat bottom, and be 7 to 10 mm (0.28 to 0.39 in) in diameter.

17 To check the valve timing, position the timing marks as specified in paragraph 5, then place the DTI and support over the second cam from the sprocket end (No 1 cylinder inlet cam).

18 Zero the DTI on the base circle of the cam (photo).

19 Carefully move the DTI, and its support if applicable, exactly 10 mm (0.394 in) towards the top of the cam. In this position the DTI should show a lift of 0.55 ± 0.05 mm (0.022 ± 0.002 in). If so, the valve timing is correct (photo).

20 If adjustment is necessary, proceed as described in paragraph 14, but working towards a correct DTI reading instead of a correct fit of the tool.

All methods

21 Check the injection pump timing (Chapter 3, Section 7), then refit the belt covers, flywheel/clutch cover, cam cover and other disturbed components.

7 Cylinder head – removal and refitting

Note: *Cylinder head bolts must not be removed from a hot engine*

1 Disconnect the battery earth lead.

2 Drain the cooling system as described in Chapter 2, Section 3. Recover the coolant if it is fit for re-use.

3 Remove the air cleaner as described in Chapter 3, Section 4.

4 Disconnect the breather hose, then unbolt and remove the camshaft cover. Also disconnect the breather hose from the camshaft carrier.

5 Remove the vacuum pump as described in Chapter 7, Section 3.

6 Disconnect the radiator top hose, the heater hose and the coolant bleed hose from the cylinder head (photo). On models where the expansion tank main hose obstructs head removal, disconnect the hose and move it aside.

7 Disconnect the coolant temperature sender lead from the thermostat housing.

8 Clean around the injection pipe unions, then unscrew them from the injectors and from the injection pump. Be prepared for fuel spillage. Disconnect the fuel return line.

9 Disconnect the glow plug wire from the bus bar.

10 Remove the alternator drivebelt and (when fitted) the steering pump drivebelt.

11 Remove the camshaft drivebelt covers and the clutch/flywheel cover, position the engine at TDC (No 1 firing) and remove the camshaft drivebelt from the camshaft sprocket. Refer to Section 5 for fuller details; if the valve timing has been satisfactory so far, make sure that the camshaft is locked or that marks are made so that the timing is not lost.

12 Unbolt the exhaust downpipe from the exhaust manifold and the flexible coupling. Release the pipe from its mounting bracket and remove it. Recover the manifold flange gasket.

13 Slacken the cylinder head bolts a quarter turn each, working in a spiral sequence from outside to inside. Follow the same sequence and slacken the bolts another half turn, then remove the bolts and washers. Obtain new bolts for reassembly.

6.18 Dial test indicator zeroed on base circle of second cam

6.19 Another reading is taken towards the top of the cam

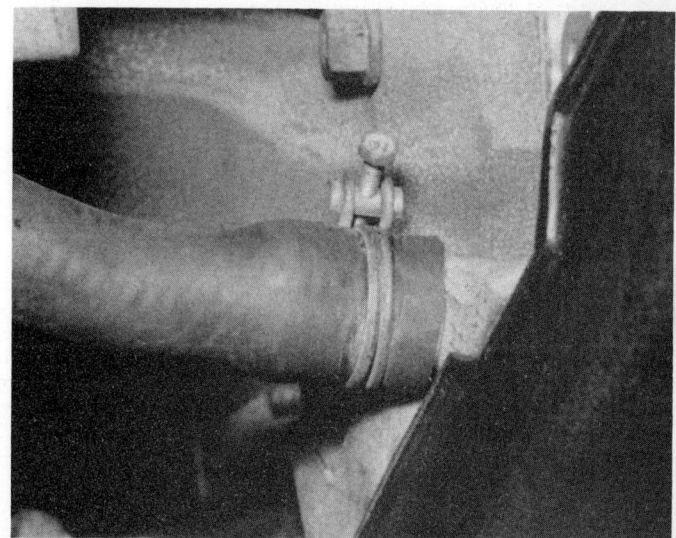

7.6 Heater hose connection to cylinder head

14 Unbolt and remove the thermostat housing, disengaging it from the coolant distribution pipe.

15 Remove the camshaft carrier and camshaft.

16 Recover the loose valve gear (cam followers, thrust pads and hydraulic lifters) from the top of the head (photos). Immerse the hydraulic lifters in a container of clean engine oil to avoid any possibility of draining. Keep all of the components in order if they are to be re-used.

17 Lift off the cylinder head and manifolds, using the manifolds as handles to rock it free if necessary. Do not prise between the mating faces to free a stuck head. A couple of lugs are provided at the ends of the block and head for prising. Recover the gasket.

18 Clean and examine the cylinder head and its components as described in Sections 22 to 24. Also clean the head mating face on the block.

19 Commence refitting by selecting the correct gasket. Three thicknesses of gasket are available; they are identified by the presence or absence of notches in the position shown (Fig. 1.6). The trusting reader may choose to fit a new gasket of the same thickness as that removed, provided that no work has been carried out on the pistons and there is no evidence of piston/valve contact. It is certainly wise, and essential if the pistons have been disturbed, to measure piston projection as described in Section 28 and then select the correct gasket. See the Specifications for piston projection and gasket thickness figures.

20 Fit the new head gaskets to the block. Make sure the gasket is the right way up – it is marked OBEN/TOP (photo). Do not use any jointing compound.

21 Place the head on the block, making sure that the dowels engage in their recesses.

22 Refit the valve lifters, thrust pads and cam followers to their original locations (if applicable), lubricating them generously with clean engine oil. If new hydraulic lifters are being used, initially immerse each one in a container of clean engine oil and compress it (by hand) several times to charge it.

23 Wipe clean the camshaft carrier and head mating faces and coat them with jointing compound to GM spec 15 03 166. Fit the camshaft and carrier to the head.

7.16A Remove the cam follower ...

7.16B ... the thrust pad (arrowed) ...

7.16C ... and the hydraulic lifter

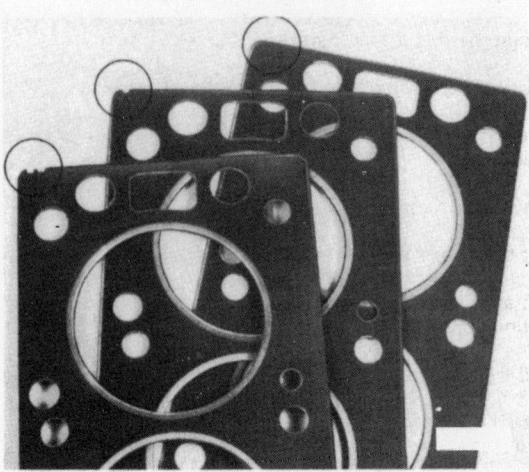

Fig. 1.6 Head gasket thickness is shown by notches (circled) (Sec 6)

Chapter 1 Engine

7.20 Head gasket OBEN/TOP marking

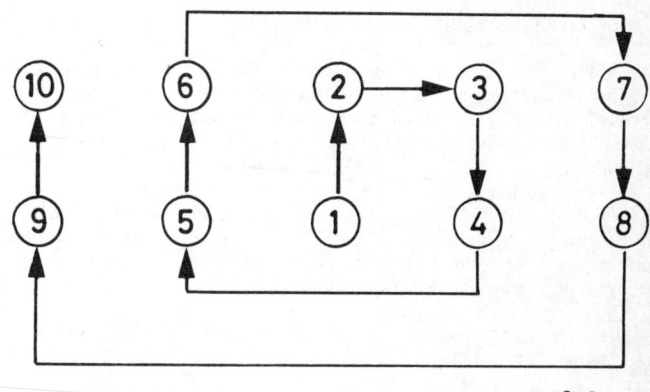

Fig. 1.7 Cylinder head bolt tightening sequence (Sec 7)

24 Fit the new head bolts with their washers. Nip up the bolts in a spiral sequence, working from the centre outwards. Make sure that the camshaft carrier is pulled down evenly.

25 In the same sequence, tighten the bolts to the Stage 1 specified torque. Subsequent tightening is by angular rotation rather than torque: use a protractor or a marked card if necessary to indicate the angles. Tighten the bolts in the same sequence through the angles specified for Stages 2, 3 and 4.

26 Refit the thermostat housing, using new O-rings on the head face and on the coolant distribution pipe.

27 Refit and tension the camshaft drivebelt as described in Section 5, then check the valve timing as described in Section 6.

28 Refit and connect the vacuum pump, using new sealing rings. Refer to Chapter 5 if necessary.

29 Refit the camshaft cover, using a new gasket, and tighten its securing bolts.

30 Refit and secure the camshaft drivebelt covers.

31 Refit and tension the alternator drivebelt and (when applicable) the steering pump drivebelt.

32 Reconnect the breather hoses and coolant hoses.

33 Reconnect the coolant temperature sender and glow plug leads.

34 Refit and secure the fuel injection pipework. Also reconnect the fuel return line.

35 Refit the air cleaner.

36 Refit the clutch/flywheel cover

37 Refit and secure the exhaust pipe, using a new gasket and a new sealing ring if necessary.

38 Refill the cooling system.

39 Check that nothing has been overlooked, then reconnect the battery and start the engine.

40 Run the engine to operating temperature, then stop it and tighten the head bolts in sequence through the angle specified for Stage 5 (photo).

7.40 Tightening a cylinder head bolt through a set angle

41 After approximately 600 miles (1000 km) have been covered, tighten the bolts through the angle specified for Stage 6. This should be done with the engine cold or warm (not hot).

8 Camshaft – removal and refitting

1 There is a special tool available (KM 2068) by means of which the camshaft can be removed without disturbing the camshaft carrier. The tool consists of a series of plates, carrying adjustable feet, which bolt to the camshaft carrier so that the feet depress the cam followers. The camshaft can then be withdrawn after removal of the drivebelt, sprocket and thrust plate. The pistons must be placed at mid-stroke (90° BTDC) for this operation.

2 In the absence of the special tool, the camshaft must be removed with the camshaft carrier. Since this entails the removal of the cylinder head bolts, strictly speaking the cylinder head should be removed and the gasket renewed. Bearing in mind the greater compression pressures in this type of engine, it is probably most unwise not to renew the gasket.

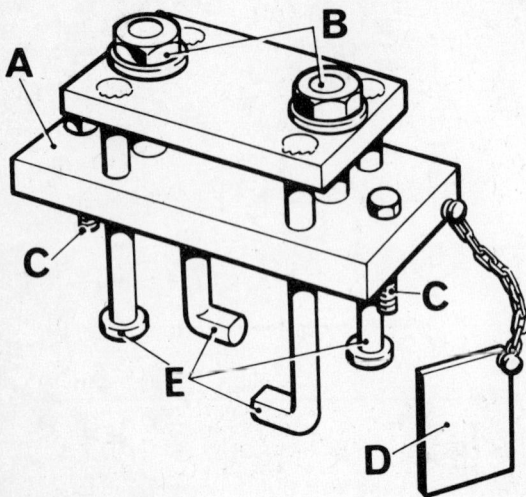

Fig. 1.8 One of the four sections of camshaft removal tool KM-2068 (Sec 8)

A Plate   D Spacer (not used)
B Nuts   E Feet
C Bolts

3 With the camshaft and carrier removed as described in the previous Section, restrain the camshaft from turning and unbolt the sprocket. Strike the sprocket lightly, if need be, to force it from its taper.

4 Unbolt and remove the camshaft thrust plate (photos).

5 Withdraw the camshaft, being careful not to catch the cam lobes on the carrier bearings (photo). Be careful also not to be injured by any sharp edges which may be present.

6 Examination of the camshaft and carrier is considered in Section 26.

7 Oil the bearing surfaces before commencing refitting.

8 Fit a new oil seal to the sprocket end of the carrier.

9 Insert the camshaft, again being careful of the bearing surfaces. Fit and secure the thrust plate, then check the camshaft endfloat (photo). Renew the thrust plate if endfloat is excessive.

10 Loosely fit the camshaft sprocket, using a new bolt, but do not tighten it yet.

11 Position the camshaft in roughly the correct position for TDC, No 1 firing – ie both cam lobes for No 1 cylinder inclined upwards, and those for No 4 cylinder inclined downwards.

12 Refit the camshaft and carrier as described in the previous Section.

13 If a new camshaft has been fitted, coat the lobes and cam followers with molybdenum disulphide paste, or with special cam lube if this was supplied with the new camshaft.

14 Unless otherwise instructed by the camshaft maker, observe the following running-in schedule immediately after start-up:

   (a) 1 minute at 2000 rpm
   (b) 1 minute at 1500 rpm
   (c) 1 minute at 3000 rpm
   (d) 1 minute at 2000 rpm

Precise speeds are not important, the main thing is that the engine should not be left to idle until this initial running-in has taken place.

15 Change the engine oil (but not the filter) approximately 600 miles (1000 km) after fitting a new camshaft.

Fig. 1.9 Camshaft removal tool KM-2068 in use (Sec 8)

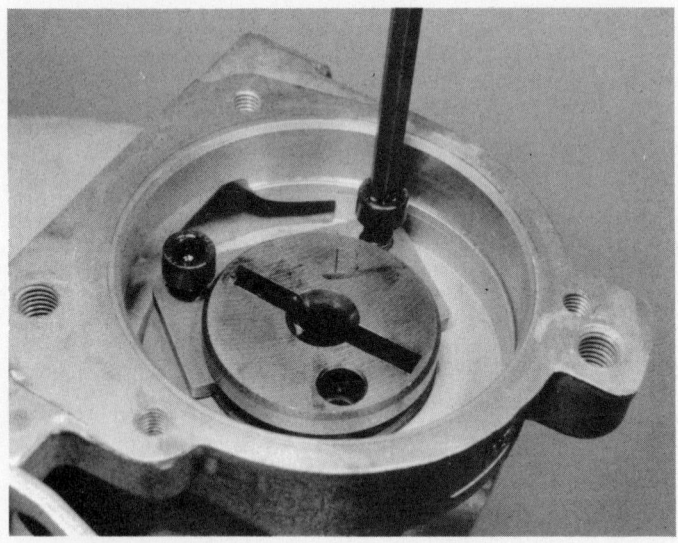

8.4A Undoing the camshaft thrust plate Allen screws

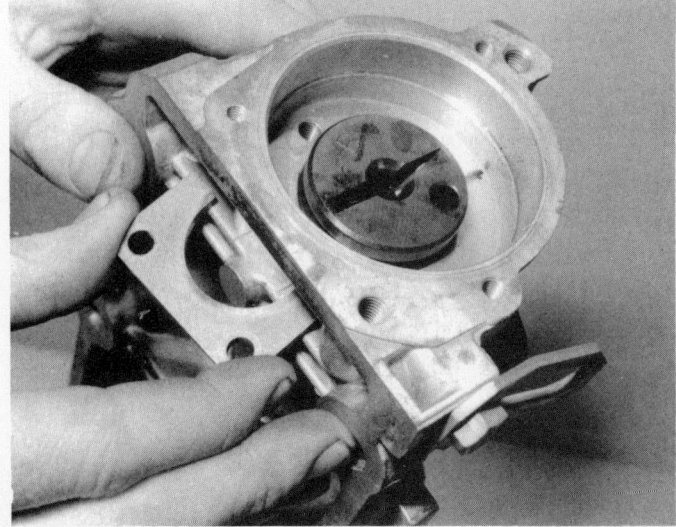

8.4B Removing the camshaft thrust plate

8.5 Removing the camshaft from the carrier

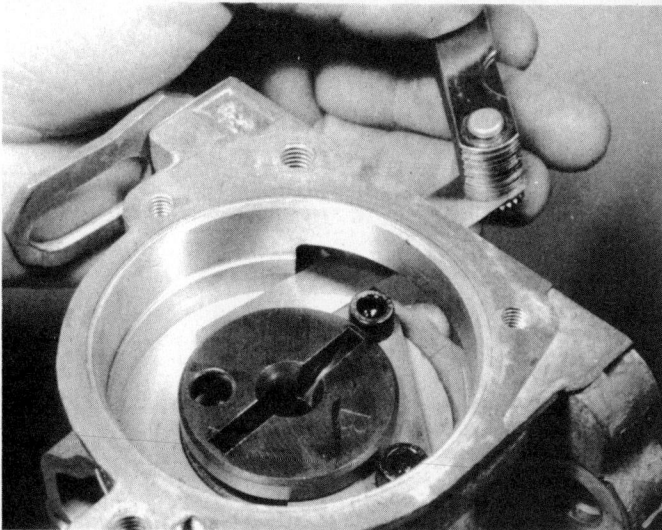

8.9 Measuring camshaft endfloat

9 Sump – removal and refitting

1 Raise and securely support the front of the vehicle, or place it over a pit.

2 Remove the sump drain plug and drain the engine oil into a suitable container. Refit the plug when draining is complete.

3 Remove the sump securing bolts – 14 on the engine examined for this book – and their special washers.

4 Pull the sump off the cylinder block. If it is stuck, strike it on the sides with a wooden or plastic mallet. Only lever between the sump and block as a last resort, as damage may be caused to the sealing faces.

5 Clean the inside of the sump. Remove all traces of gasket from the sump and block faces.

6 Commence refitting by applying RTV sealant (to GM spec 15 03 294) to the joints between the rear main bearing cap and the block, and the oil pump and the block.

7 Smear the new sump gasket with grease to improve its adhesion and position it on the block or sump.

8 Offer the sump to the block, making sure that the gasket is not displaced. Secure with the bolts and washers, using thread locking compound on the bolt threads.

9 Tighten the bolts in an even sequence to the specified torque.

10 Make sure that the drain plug is fitted and tightened, then refill the engine with oil.

10 Oil pump – removal and refitting

1 Remove the sump as described in the previous Section.

2 Remove the oil filter as described in Section 4. Also disconnect the oil pressure warning light switch.

3 Bring the engine to TDC, No 1 firing, then slacken the camshaft drivebelt as described in Section 5.

4 Jam the starter ring gear and unscrew the crankshaft pulley/ sprocket centre bolt. This bolt is extremely tight. Remove the bolt and the pulley/sprocket assembly; recover the Woodruff key.

5 Remove the drivebelt idler pulley, which is secured by a large Allen screw through its hub.

6 Remove the drivebelt backplate – this is secured by some of the drivebelt cover screws.

7 Remove the six Allen screws which secure the oil pump to the block, and the single bolt which secures the oil pump pick-up tube to the block lower face. Note that this latter bolt appears identical to the sump bolts – keep it separate or it may be considered 'lost' until after the sump has been refitted.

8 Pull the pump off its dowels and remove it with the pick-up tube. Recover the gasket and unbolt the pick-up tube.

9 Overhaul of the oil pump is described in Section 21.

10 Commence refitting by lubricating the lips of the oil seal.

11 Fit the pump to the block, using a new gasket. Be careful not to damage the oil seal on the step on the crankshaft nose. Fit and tighten the securing screws.

12 Refit and secure the pick-up tube, using a new O-ring at the pump flange.

13 Refit the drivebelt backplate and the idler pulleys. Tighten the idler pulley Allen screw.

14 Refit the Woodruff key to the crankshaft nose.

15 Engage the sprocket with the drivebelt and fit the sprocket/pulley assembly to the crankshaft nose.

16 Fit the sprocket/pulley centre bolt. Jam the ring gear teeth and tighten the bolt to the specified torque.

17 Tension the camshaft drivebelt as described in Section 5, then check the valve timing as described in Section 6.

18 Refit the sump as described in Section 9.

11 Pistons and connecting rods – removal and refitting

1 Remove the cylinder head as described in Section 7.

Chapter 1 Engine

2 Remove the sump as described in Section 9.

3 Examine the connecting rods and caps for identification and match marks. If none are present, make paint or punch marks to identify the rods and caps and to make sure that the caps are refitted the right way round.

4 Bring two pistons to BDC. Remove the connecting rod cap nuts, take off the caps and push the pistons up through the top of the block. Recover the bearing shells, keeping them with their original rods if they are to be re-used.

5 If there is a pronounced wear ridge at the top of the bore, the piston or its rings may be damaged during removal. This can be avoided by removing the ridge with a tool known as a ridge reamer. (Such a degree of bore wear would probably mean that a rebore was necessary, which would also entail the fitting of new pistons in any case.)

6 Bring the other two pistons to BDC and remove them.

7 If it is proposed to separate the pistons from the connecting rods, make identifying marks on the piston crowns first so that they can be refitted to their original rods and the right way round.

8 Remove the circlip from the gudgeon pin bore on one side of the piston. From the other side, press out the gudgeon pin by hand. No great force should be required (photos).

9 For examination of the pistons and rings, refer to Section 26.

10 When refitting the pistons to the connecting rods, oil the gudgeon pins liberally first. Push the pins home and secure them with the circlips. Some patience may be needed when fitting the circlips: they are liable to spring out.

11 If new pistons or rings are to be fitted to old bores, it is essential to deglaze the bores to allow the new rings to bed in. Do this using medium grit emery paper, or a flap wheel in an electric drill, with a circular up-and-down action. The aim is to produce a criss-cross pattern on the bores. Protect the bearing journals on the crankshaft by covering them with masking tape during this operation. Clean the bores thoroughly with a paraffin-soaked rag and dry them with a clean cloth. Then remove the masking tape from the journals and clean them.

12 Commence refitting by laying out the pistons, connecting rods, caps and shells in order (photo). If the piston rings have been removed, refit them and stagger the gaps as described in Section 28, paragraphs 14 to 16.

13 Make sure that the seats for the bearing shells are absolutely clean, then press the shells into their seats in the rods and caps, aligning the locating tangs correctly.

14 Lubricate the bores and the piston rings with clean engine oil. Fit a piston ring compressor to the first piston to be fitted. Turn the crankshaft to bring the journal for that piston to BDC.

15 Insert the rod and piston into the block so that the base of the ring compressor rests on the block. Make sure that the piston is the right way round (arrow on crown pointing towards the pulley end) (photo) and that the connecting rod is also the right way round (refer to the marks made when dismantling, or see Fig. 1.10).

16 Drive the piston into the bore by tapping the piston crown with, for example, the wooden handle of a hammer. If the piston does not want to go in, do not use more force, but release the compressor and see if a ring has jammed. Remove the compressor as the piston enters the bore.

17 Lubricate the crankpin and guide the connecting rod onto it, pushing the piston down the bore. Take care that the protruding studs do not scratch the crankpin, and check that the bearing shell is not displaced.

11.8A Removing a gudgeon pin circlip

11.8B Removing a gudgeon pin

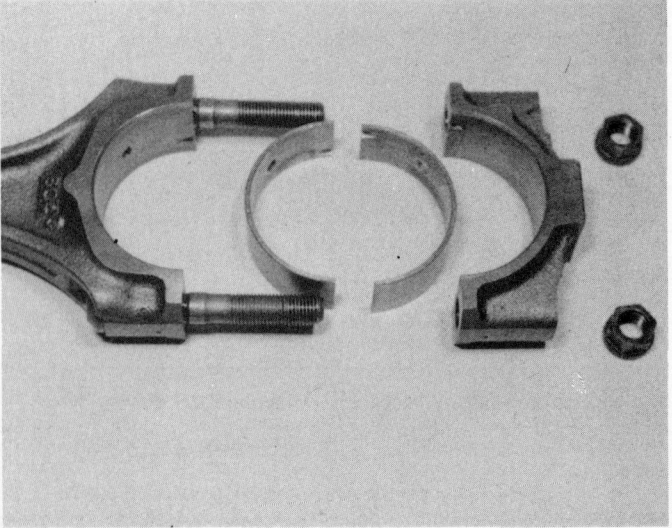

11.12 Connecting rod, bearing shells, cap and nuts

Chapter 1 Engine

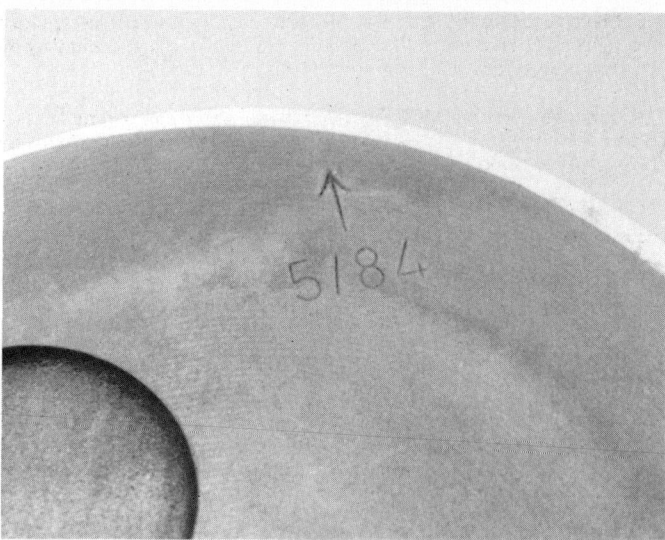

11.15 Arrow on piston crown points to pulley end

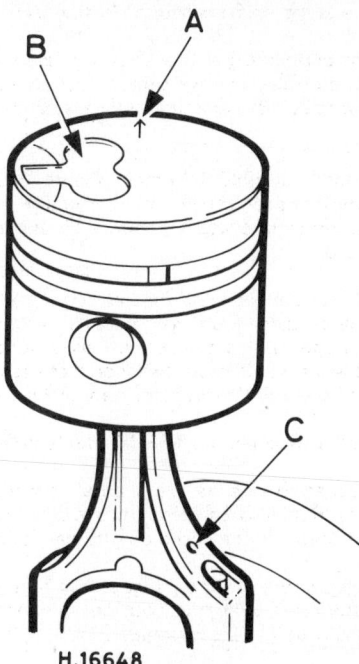

Fig. 1.10 Correct relationship between piston and connecting rod (Sec 11)

- A Arrow
- B Combustion recess
- C Oil squirt hole

18 Fit the connecting rod cap, again making sure that the shell is not displaced. Fit the nuts and tighten them to the specified torque.

19 Repeat the operations on the other pistons and rods.

20 Refit the sump and the cylinder head as described in Sections 9 and 7.

12 Flywheel – removal and refitting

This procedure applies to vehicles with manual transmission. To gain access to the driveplate on automatic transmission vehicles, the engine and transmission must first be separated; the procedure is then similar.

1 Remove the clutch as described in the appropriate manual for petrol-engined vehicles.

2 Jam the flywheel ring gear and slacken the flywheel bolts. The bolts are tight and their heads are shallow; a socket spanner without any chamfer at the business end will give the best purchase. Mark the location of the shouldered bolt relative to the flywheel and to the crankshaft flange (photo).

3 Remove the bolts and lift away the flywheel. Do not drop it, it is heavy. Obtain new bolts for reassembly.

4 Examination of the flywheel and ring gear is considered in Section 26.

5 When refitting the flywheel, line up the bolt hole for the shouldered bolt in the crankshaft flange and in the flywheel. There is no locating dowel.

6 Apply thread locking compound to the threads of the new bolts. Fit the bolts, making sure that the shouldered bolt is fitted to the correct hole. Tighten the bolts progressively to the specified torque.

7 Refit the clutch.

12.2 Shouldered bolt (arrowed) determines flywheel fitted position

13 Oil seals – renewal

Camshaft front oil seal

1 Slacken the camshaft drivebelt and remove it from the camshaft sprocket – see Section 5.

2 Remove the air cleaner as described in Chapter 3, Section 4. Disconnect the breather hose and remove the camshaft cover.

3 Hold the camshaft with an open-ended spanner on the flats provided. Unscrew and remove the camshaft sprocket bolt; obtain a new bolt for use when refitting.

4 Remove the camshaft sprocket. Tap it lightly if necessary to break the taper.

5 Carefully drill or punch a hole in the oil seal face. Screw in a self-tapping screw and use this to pull out the seal using pincers, a claw hammer or a nail bar.

6 Fill the lips of the new seal with grease. Clean out the seal seat, then press the seal into position, lips inwards. Take care not to damage the seal lips. Use a piece of tube or a socket to seat the seal.

7 Refit the camshaft sprocket and secure it with a new bolt, tightened to the specified torque.

8 Refit and tension the camshaft drivebelt as described in Section 5, then check the valve timing as described in Section 6.

9 Refit the camshaft cover and reconnect the breather hose. Refit the air cleaner as described in Chapter 3, Section 4.

Crankshaft front oil seal

10 Remove the crankshaft pulley, sprocket and Woodruff key, as described in Section 10, paragraphs 3 and 4.

11 Extract the oil seal as described in paragraph 5 of this Section.

12 Fill the lips of the new seal with grease and clean out the seal seat. Cover the step on the crankshaft nose with adhesive tape, then press the seal into position, lips inwards. Use a piece of tube or a socket to seat the seal, then remove the adhesive tape.

13 Refit the Woodruff key. Engage the sprocket with the drivebelt, fit the pulley/sprocket assembly to the crankshaft and tighten the centre bolt to the specified torque.

14 Tension the camshaft drivebelt as described in Section 5, then check the valve timing as described in Section 6.

Crankshaft rear oil seal

15 Unless various specials tools are available, the engine must be removed and separated from the transmission for this operation.

16 Remove the clutch and flywheel (or driveplate, on automatic transmission models). Obtain new bolts for reassembly.

17 Extract the oil seal as described in paragraph 5 of this Section.

18 Fill the lips of the new seal with grease and clean out the seal seat. Cut a piece of thin flexible plastic to fit around the crankshaft flange and slide the seal into place over this (photo). The seal lips face inwards.

19 Seat the seal by tapping it carefully with a wooden or plastic hammer. Make sure it seats evenly. Withdraw the plastic collar.

20 Refit the flywheel or driveplate and secure with new bolts. Refer to Section 12 for further details.

21 Refit the clutch (when applicable) and refit the engine and/or transmission to the vehicle.

Valve stem oil seals
22 Refer to Section 22.

14 Engine/transmission mountings – renewal

Ascona/Cavalier

1 The flexible mountings can be renewed if they have deteriorated. Take the weight of the engine/transmission on a hoist, or use a jack and a wooden block from below. Only remove and refit one mounting at a time.

2 Unbolt the mounting brackets from the engine or transmission and from the bodyframe (photos). Separate the flexible component from the brackets.

3 Fit the new flexible component and refit the mounting. Only nip up the bolts at first, then tighten them to the specified torque.

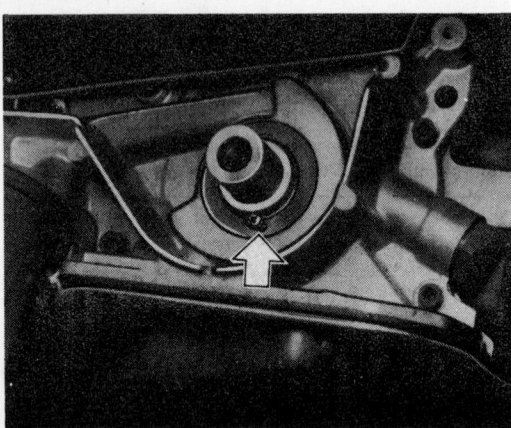

Fig. 1.11 Self-tapping screw (arrowed) screwed into crankshaft front oil seal (Sec 13)

13.18 Using a plastic collar as an aid to fitting the crankshaft rear oil seal

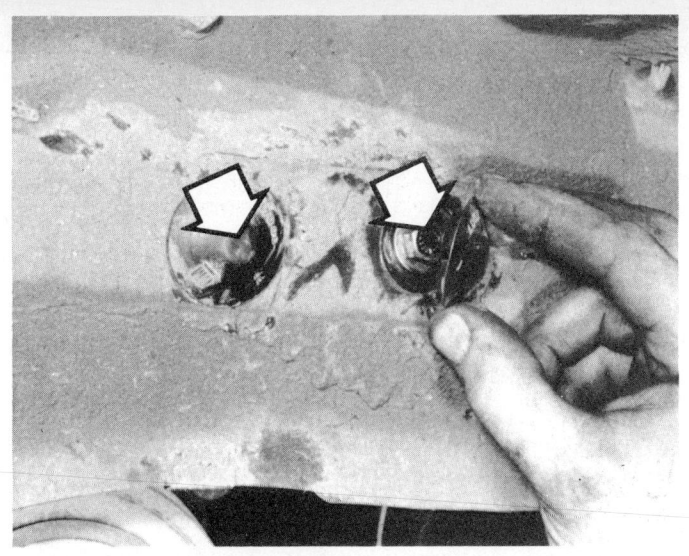

14.2A Engine mounting bolt locations – RH front (arrowed) under wheel arch

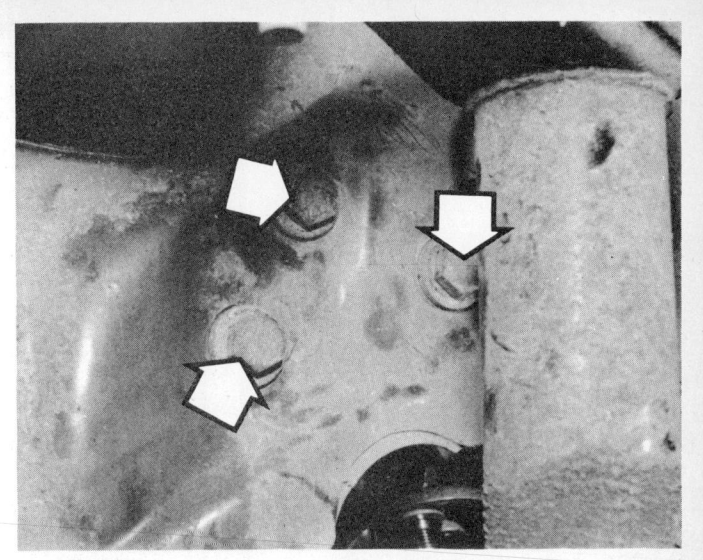

14.2B Engine mounting bolt locations – RH rear (arrowed) under wheel arch

14.2C Engine right-hand rear mounting seen from below. Alternator is normally mounted on top lug (arrowed)

14.2D Transmission mounting – LH rear seen from above

14.2E Transmission mounting – LH front seen from below

38　　　　　　　　　　　　　　　　　　　Chapter 1 Engine

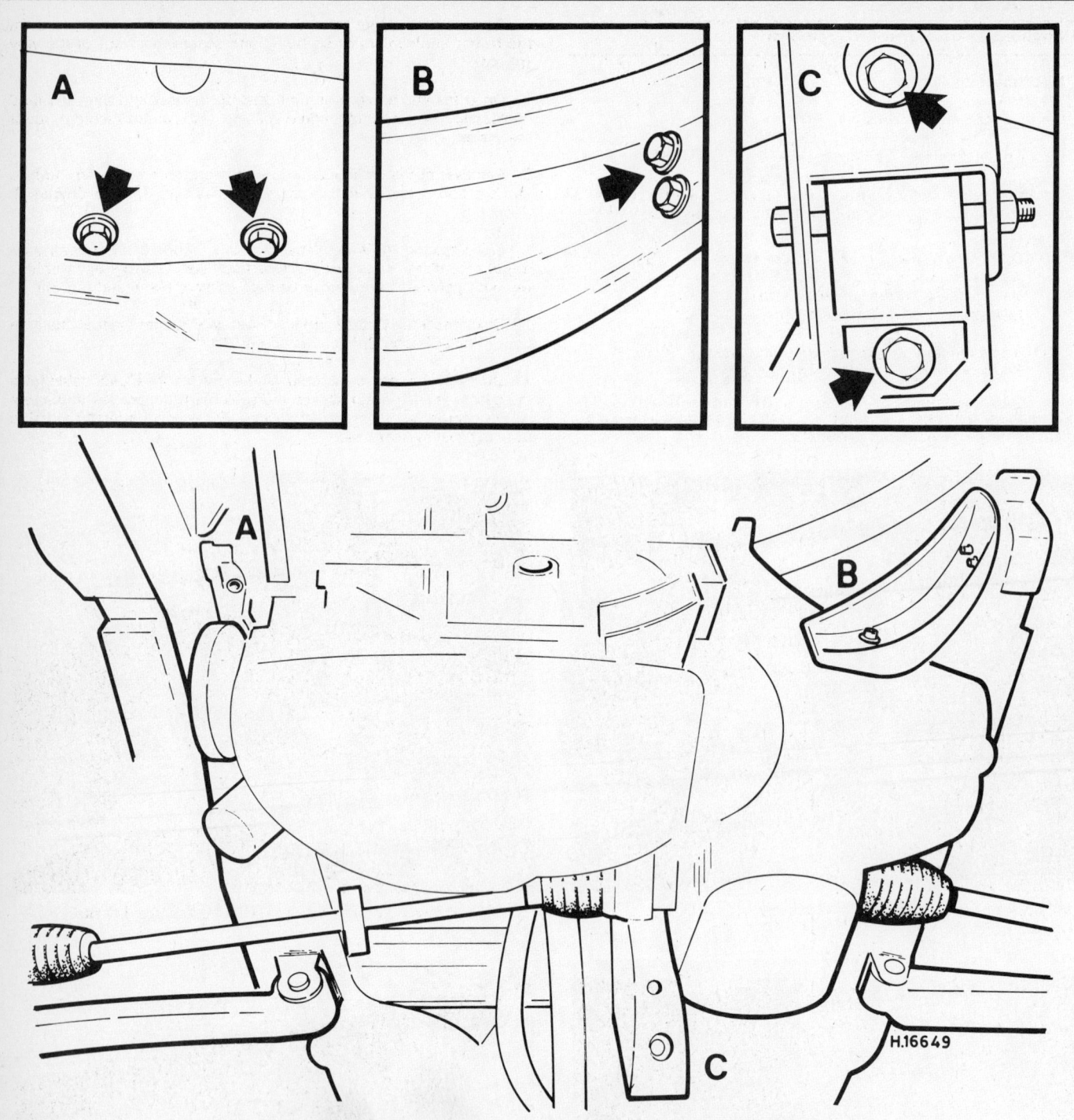

Fig. 1.12 Engine/transmission mountings – Astra/Kadett (Sec 14)

A Right-hand front　　　B Left-hand front　　　C Rear

4 Lower the hoist or jack and check that the mounting is not under strain; slacken and retighten the bolts if necessary.

Astra/Kadett/Belmont

5 The engine and transmission must be removed in order to renew the right-hand mounting. The left-hand mountings can be renewed as outlined above.

15 Engine – methods of removal

1 The engine and transmission must be removed together from below, as the engine cannot be drawn off the transmission in situ. It may be possible to remove the transmission first and then lift out the engine; but this method is untested and in any case entails extra work.

2 Engine lifting gear of adequate capacity will be required. To provide clearance to remove the engine/transmission unit from under the car, the use of a pit or a vehicle lift is desirable. Failing this, high-lift ramps or axle stands may suffice, or the front of the car may be lifted after the power unit has been lowered to the ground.

3 The procedures described in the next Section and in Section 30 relate to one particular vehicle; allowance must be made for individual model and year variations.

Chapter 1 Engine

16 Engine and transmission – removal

1 Disconnect the battery earth lead.

2 Remove the bonnet, if wished.

3 Remove the air cleaner and its snorkel.

4 Disconnect the following electrial services, making notes or attaching labels if there is any possibility of confusion later:

   (a) *Oil pressure switch (next to the oil filter)*
   (b) *Alternator (large and small leads)*
   (c) *Coolant temperature sender (photo)*
   (d) *Idle stop solenoid (on fuel pump)*
   (e) *Glow plug feed to bus bar*
   (f) *Reversing light switch (manual transmission) (photo)*
   (g) *Radiator fan and thermoswitch*

5 Drain the cooling system, taking precautions against scalding if the coolant is hot. See Chapter 2, Section 3.

6 Remove the radiator top and bottom hoses completely. Disconnect the heater and coolant bleed hoses and secure them out of the way (photo).

7 On automatic transmission models, disconnect and plug the fluid cooler lines. Be prepared for fluid spillage; take care to keep dirt out of the system.

8 Removal of the radiator and cooling fan is recommended, both to improve access and to reduce the risk of damage. Refer to Chapter 2, Section 6.

9 Clean around the fuel inlet and return unions at the pump, then disconnect them. Be prepared for fuel spillage. Cap the open unions (eg with polythene and rubber bands) to keep fuel in and dirt out.

10 Disconnect the throttle and cold start cables from the fuel injection pump – see Chapter 3, Sections 13 and 14.

11 On manual transmission models, disconnect the clutch cable from the release lever (photo). Unclip the cable from the bracket at the rear of the engine; note the insulating material which protects the cable from exhaust system heat.

16.4A Coolant temperature sender lead (arrowed) next to thermostat housing

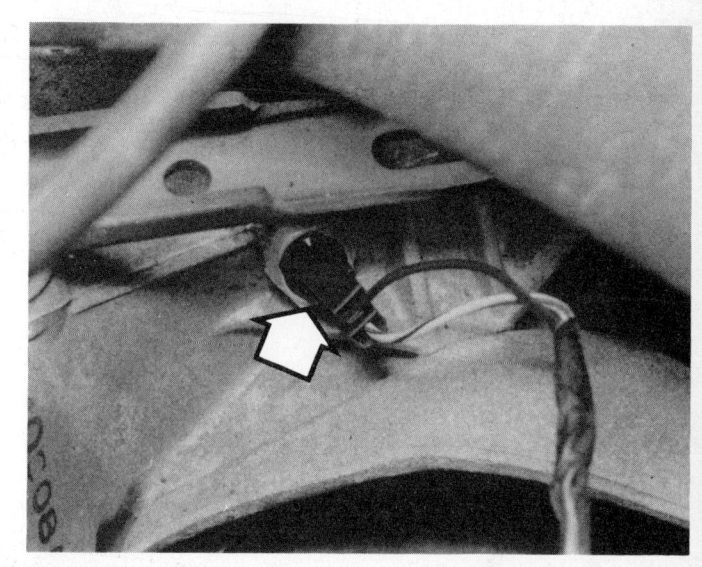

16.4B Reversing light switch (arrowed) on transmission case

16.6 Heater hose and radiator bottom hose connections to large coolant pipe

16.11 Clutch cable attachment to release lever

Chapter 1 Engine

12 On manual transmission models, disconnect the gearshift linkage just behind the universal joint.

13 On automatic transmission models, disconnect the gearshift and kickdown cables.

14 On all models, disconnect the speedometer cable (photo).

15 Disconnect the starter motor supply lead from the battery positive terminal. Separate the starter command lead connector at the same point (photo).

16 Unbolt the exhaust downpipe from the manifold flange and the flexible joint. Free the pipe from any intermediate brackets and remove it. Recover the flange gasket (photos).

17 Release the clip which secures the fuel hoses to the camshaft drivebelt cover. Tie the hoses out of the way.

18 Remove the power steering pump, when fitted, and secure its pipes out of the way.

19 Disconnect the servo vacuum pipe from the vacuum pump.

20 Raise and support the front of the vehicle and remove the front wheels.

21 Separate the front suspension bottom balljoints (photo).

22 Release the driveshafts from the transmission – refer to the appropriate manual for petrol-engined vehicles. Be prepared for oil spillage. Secure the driveshafts out of the way.

23 Disconnect the earth strap from the transmission end cover (photo).

24 Take the weight of the engine/transmission on the lifting gear, using the lifting eyes provided.

25 Unbolt the right-hand engine mounting from the top (two bolts) and side (two bolts accessible through the wheel arch) – see Section 14.

26 Unbolt and remove the other engine/transmission mounting brackets.

27 On models where the alternator mounting forms part of an engine

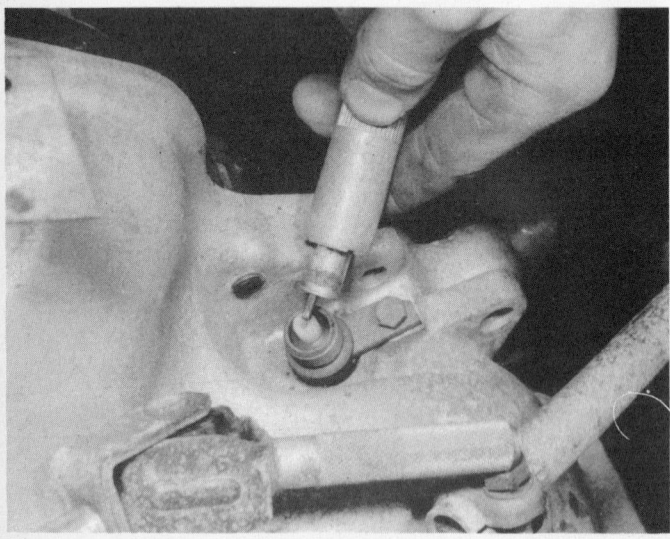

16.14 Disconnecting the speedometer cable (manual transmission). Separated gearshift linkage is in foreground

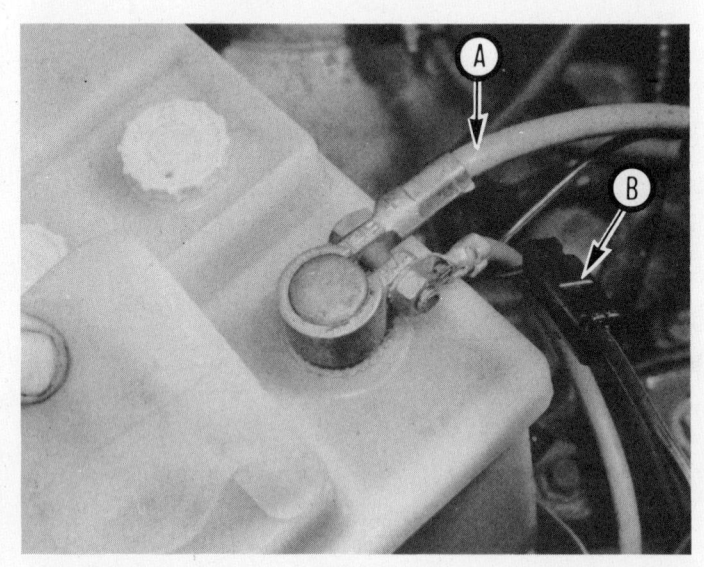

16.15 Starter motor supply lead (A) and command lead connector (B)

16.16A Exhaust system flexible joint

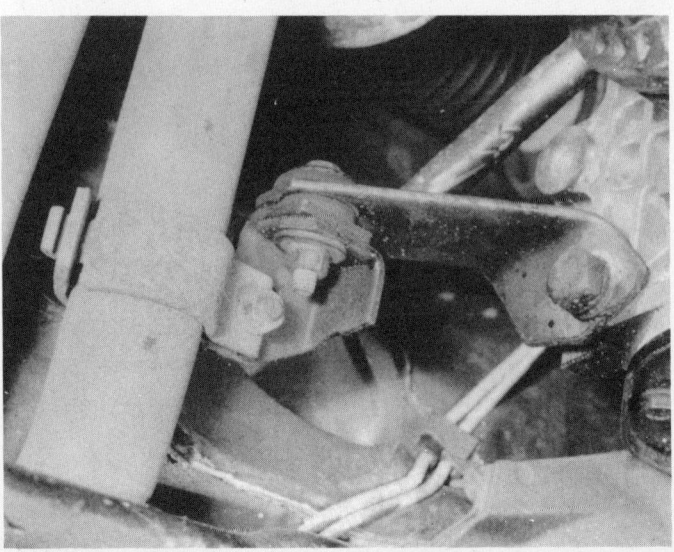

16.16B Exhaust downpipe-to-transmission bracket

Chapter 1 Engine

16.21 Front suspension bottom balljoint separated

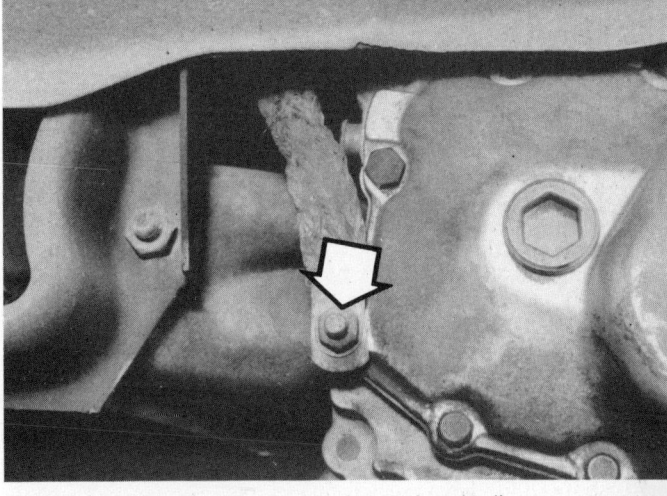

16.23 Transmission earth strap attachment (arrowed)

mounting bracket, either remove the alternator or secure it so that it cannot fall off.

28 If there is limited side-to-side clearance for lowering the engine/transmission unit, removing the oil filter and the crankshaft pulley will improve matters slightly.

29 Check that no attachments have been overlooked, then lower the engine/transmission to the ground. Withdraw the unit from under the vehicle and take it to the bench.

17 Engine and transmission – separation

1 Clean the outside of the engine and transmission.

2 Unbolt and remove the starter motor and its heat shield.

3 Unbolt and remove the clutch or torque converter cover plate from the bottom of the bellhousing.

4 On automatic transmission models, unbolt the torque converter from the driveplate. Turn the crankshaft to gain access to the bolts through the cover plate aperture. Obtain new bolts for reassembly.

5 On all models, bring the engine to TDC, No 1 firing. (With the transmission removed there will be no TDC indicated).

6 Support the transmission and remove the engine-to-transmission bolts.

7 Withdraw the transmission from the engine. Do not allow the weight of the transmission to hang on the input shaft. With automatic transmission, make sure that the torque converter stays in its housing.

8 Make alternative TDC marks between the flywheel and the engine block for use during dismantling and reassembly.

18 Engine dismantling – general

1 If the engine has been removed from the car for major overhaul or if individual components have been removed for repair or renewal, observe the following general hints on dismantling and reassembly.

2 Drain the oil into a suitable container and then thoroughly clean the exterior of the engine using a degreasing solvent or paraffin. Clean

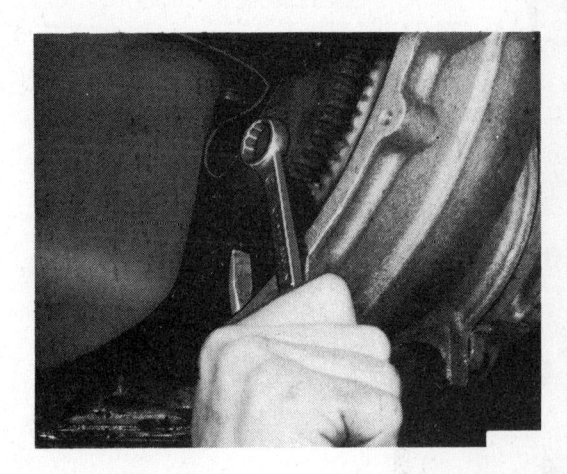

Fig. 1.13 Removing a torque converter-to-driveplate bolt (Sec 17)

away as much of the external dirt and grease as possible before dismantling.

3 As parts are removed, clean them in a paraffin bath. However, do not immerse parts with internal oilways in paraffin as it is difficult to remove, usually requiring a high pressure hose. Clean oilways with nylon pipe cleaners.

4 Avoid working with the engine or any of the components directly on a concrete floor, as grit presents a real source of trouble.

5 Wherever possible work should be carried out with the engine or individual components on a strong bench. If the work must be done on the floor, cover it with a board or sheets of newspaper.

6 Have plenty of clean, lint-free rags available and also some containers or trays to hold small items. This will help during reassembly and also prevent possible losses.

7 Always obtain a complete set of new gaskets if the engine is being completely dismantled, or all those necessary for the individual component or assembly being worked on. Keep the old gaskets with a view to using them as a pattern to make a replacement if a new one is not available.

19 Ancillary components – removal

1 If the engine is to be completely dismantled, remove the following ancillary components (if not already done):

    (a) Alternator and drivebelt
    (b) Inlet and exhaust manifolds (Chapter 3)
    (c) Camshaft drivebelt (Section 5)
    (d) Water pump and thermostat (Chapter 2)
    (e) Fuel injection pump (Chapter 3) and its bracket
    (f) Vacuum pump (Chapter 5)
    (g) Clutch assembly (Chapter 4)
    (h) Fuel injectors and glow plugs (Chapter 3)
    (j) Oil filter (Section 4)
    (k) Dipstick, dipstick tube and cable bracket (photo)

2 Also remove the large coolant pipe which runs along the rear face of the engine, and the hose which connects it to the water pump (photo).

20 Engine – complete dismantling

1 Remove the following items as described in the appropriate Sections, disregarding those instructions which have already been carried out in removing the engine:

    (a) Cylinder head (Section 7)
    (b) Sump (Section 9) – if possible, without inverting the engine, so that any sludge does not enter the crankcase
    (c) Oil pump (Section 10)
    (d) Pistons and connecting rods (Section 11)
    (e) Flywheel or driveplate (Section 12)

2 Invert the engine so that it is standing on the top face of the block.

3 Check that the main bearing caps are numbered 1 to 4 from the pulley end. (The fifth cap is not numbered, but it is different from the rest). The numbers are on the rear (water pump) halves of the caps. Make identification marks if no numbers are present.

4 Remove the bolts which secure the main bearing caps (photo).

5 Lift off the main bearing caps; tap them lightly to free them if they are stuck. Keep the bearing shells with their caps if they are to be re-used. Clean all traces of sealant from the rear cap.

6 Lift out the crankshaft and put it somewhere safe. Do not drop it, it is heavy.

7 Extract the upper half main bearing shells. Keep them in order if they are to be re-used.

8 With the exception of semi-permanent items such as dowels and core plugs, dismantling of the engine is complete.

21 Oil pump – dismantling, overhaul and reassembly

1 With the oil pump removed from the vehicle, withdraw the rear cover. The cross-head fixing screws are very tight and an impact driver will be required to remove them (photo).

2 Check the clearance between the inner and outer gear teeth and the outer gear and the pump body (photos).

3 Using a straight-edge across the pump cover flange, measure the gear endfloat (photo).

4 If any of the clearances are outside the specified tolerance, renew the components as necessary. Note that the outer gear face is marked for position (photo).

19.1 Dipstick tube and cable bracket securing bolts (arrowed)

19.2 Two bolts (arrowed) securing large coolant pipe

20.4 Removing a main bearing cap bolt

Chapter 1 Engine 43

21.1 Removing the oil pump rear cover screws

21.2A Oil pump clearance check – inner-to-outer teeth

21.2B Oil pump clearance check – outer gear-to-body

21.3 Oil pump gear endfloat check

21.4 Oil pump outer gear position marking (arrowed) must face rearwards

5 The pressure regulator valve can be unscrewed from the oil pump housing and the components cleaned and examined (photo).

6 The oil pressure warning light switch is screwed into the rear of the pump (photo). If it is suspected of being defective it should be renewed.

7 The oil filter bypass valve can be removed from the filter carrier by screwing an M10 tap into it (photo). Drive the new valve into position up to its stop.

8 Before reassembling the pump, take the opportunity to renew the front oil seal.

9 Refit the gears, lubricating them with clean engine oil.

10 Refit the rear cover (no gasket is used). Insert and tighten the retaining screws.

11 Lubricate the pressure regulator valve plunger with clean engine oil. Refit the plunger and spring and secure them with the plug and washer. Tighten the plug to the specified torque.

44 Chapter 1 Engine

12 Refit and tighten the oil pressure warning light switch, if it was removed.

22 Cylinder head – dismantling and reassembly

1 If not already done, remove the fuel injectors and the glow plugs – see Chapter 3, Sections 10 and 12.

2 A valve spring compressor will be needed to remove the valve and spring. Fit the compressor foot to the head of a valve, with the forked end over the valve spring cap. Operate the compressor so that the cap is depressed a little way and extract the split collets (photo). If the cap seems to be stuck, remove the compressor and strike the cap smartly with a mallet, then try again.

3 With the collets removed, release the compressor and remove the spring cap, spring and valve rotator. Withdraw the valve from the other side of the head.

4 Remove the valve stem oil seal from the top of the valve guide. New seals will be needed for reassembly.

5 Repeat these operations to remove the remaining valves. Keep each valve and its associated components separate, in a segmented box.

6 The swirl chambers can be driven out if necessary using a non-ferrous drift through the injector hole (photo). The injector insulating sleeves can then be driven out in the opposite direction, using a non-ferrous drift 10 mm (0.39 in) in diameter. Recover the sleeve sealing rings.

7 There is an oil pressure relief valve in the cylinder head to protect the valve hydraulic lifters (photo). Renewal of this valve is best left to a GM dealer, since various special tools are required to extract the valve cage and seat. With the old valve removed it is simple enough to drive a new one into place through the core plug aperture, then seal with a new plug.

8 Thoroughly clean the cylinder head. Examine it and its components as described in Section 23.

9 If the swirl chambers and insulating sleeves were removed, the swirl chambers should be refitted first. Make sure that the locating ball is correctly positioned (Fig. 1.16). From the other side of the head fit new sealing rings for the insulating sleeves, then drive them into position.

21.5 Oil pump pressure regulator valve components

21.6 Oil pressure warning light switch

21.7 Oil filter bypass valve (arrowed)

22.2 Compress the valve spring and remove the collets (arrowed)

Chapter 1 Engine

45

22.6 Driving out a swirl chamber

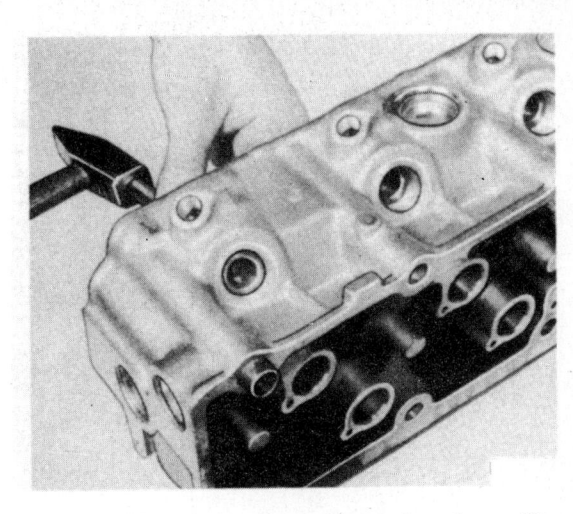

Fig. 1.14 Driving out an injector insulating sleeve (Sec 22)

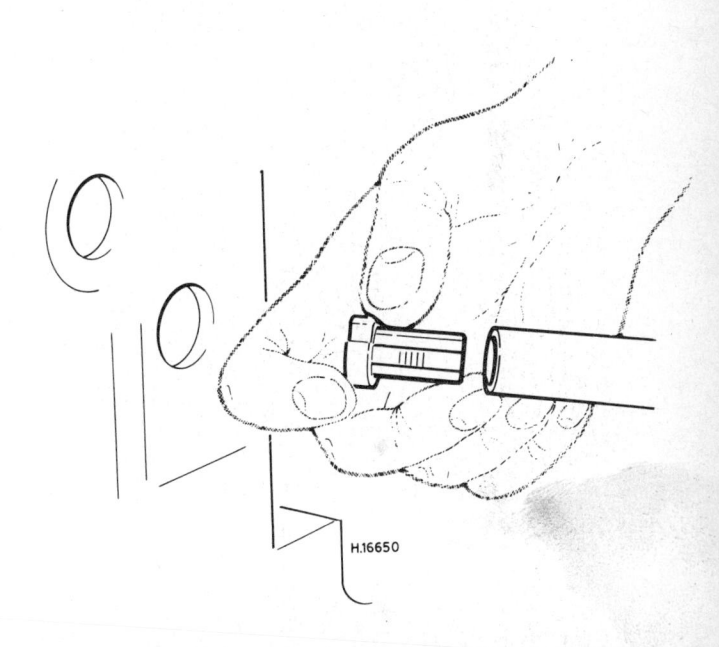

22.7 Oil pressure relief valve (arrowed) in cylinder head

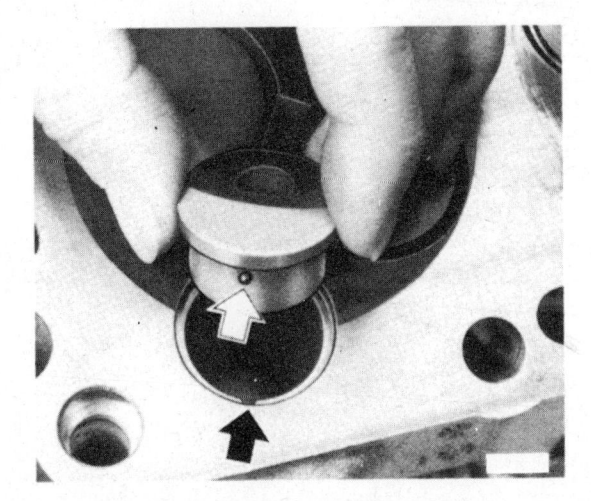

Fig. 1.15 Fitting a cylinder head oil pressure relief valve (Sec 22)

Fig. 1.16 Fitting a swirl chamber. Note locating ball and recess (arrows) (Sec 22)

10 After refitting the swirl chambers, check their projection as described in Section 23.

11 Lubricate a valve stem with clean engine oil and fit the valve into its guide (photo).

12 From the other side of the head, fit the valve rotator (photo).

13 A temporary protective sleeve should have been supplied with the new valve stem oil seals. Fit the sleeve over the valve stem and lubricate it, then press the seal into position over the sleeve (photos). Use a piece of tube or a small box spanner if necessary to push the seal onto its seat. The outer part of the seal engages with the groove on the valve guide. Remove the sleeve.

14 Fit the valve spring (either way up) and the spring cap (photos).

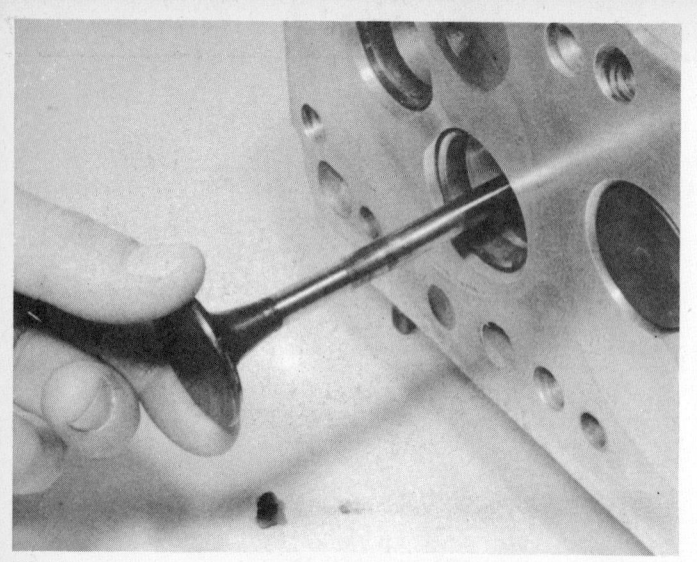

22.11 Inserting a valve into its guide

22.12 Valve rotator in position

22.13A Fit and lubricate the protective sleeve ...

22.13B ... then fit the valve stem oil seal over the sleeve

22.14A Fit the valve spring ...

22.14B ... and the valve spring cap

15 Fit the valve spring compressor and depress the spring cap far enough to allow the collets to be fitted. A smear of grease on the valve stem will keep them in position. Carefully release the compressor, making sure that the collets are not displaced. Tap the spring cap with a mallet to settle the components.

16 Refit the other valves, springs and associated components in a similiar fashion.

17 Check the valve head recess as described in Section 23.

18 Refit the fuel injectors and the glow plugs as described in Chapter 3.

23 Cylinder head – examination and overhaul

1 Check the head visually for cracks or other obvious signs of damage. (Small cracks between the inlet and exhaust valve seats are not serious and may be ignored).

2 Examine the valve sealing faces and the valve seats in the head for pitting, burning or other damage. Light pitting may be rectified by grinding; anything more serious means that the valve and/or seat must be refaced or renewed. Consult a GM dealer or other specialist if work of this nature appears necessary.

3 Check the fit of the valves in their guides. Worn guides must be reamed out to the next oversize – another specialist job – and valves with oversize stems fitted. Permissible oversizes are listed in the Specifications. Note that oversize valves may already have been fitted – look for grade markings on the valve guide and valve stem. Check also that the valve stems are not bent.

4 Valve grinding is carried out as follows. Smear a little coarse grinding paste onto the sealing area of the valve head. Insert the valve into its guide and using a suitable tool, typically a stick with a rubber sucker on it, grind the valve to its seat with a semi-rotary motion. Lift the valve occasionally to redistribute the grinding paste. Do not use an electric drill or other tool with a continuous circular motion.

5 When an unbroken ring of grinding paste is present on the valve head and its seat, wipe off the coarse paste and repeat the process with fine paste. If the valve seats have been recut accurately, and new or refaced valves are being fitted, only fine paste need be used from the outset.

Fig. 1.17 Grinding a valve to its seat (Sec 23)

6 When valve grinding is complete, clean away all traces of grinding paste using first a paraffin-soaked rag, then a clean dry rag, then (if available) an air line. Any grinding paste remaining may cause rapid wear.

7 Renew the valve springs if they are obviously distorted or weak, or if they have seen much service. Springs do fatigue in time and it is better to renew them now than to have one break in use.

8 Check the head mating faces, both block and camshaft carrier, for freedom from distortion. Use a straight-edge and feeler blades; do not be deceived by the (permitted) projection of the swirl chambers. Distortion limits are given in the Specifications; an out-of-limit head may be reclaimed by specialist machining, provided that the overall height of the head is not reduced below that specified, and that valve recess and swirl chamber projection stay within limits.

9 Check the projection of the swirl chambers from the head-to-block face, using a straight-edge and feeler blades or (preferably, for such a small measurement) a dial test indicator (photos). The allowable projection is given in the Specifications.

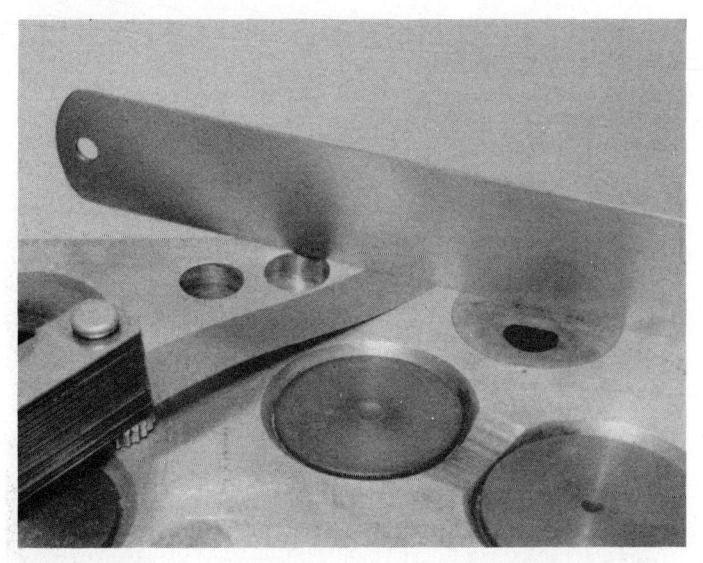

23.9A Checking swirl chamber projection with a straight-edge and feeler blades

23.9B Checking swirl chamber projection with a dial test indicator

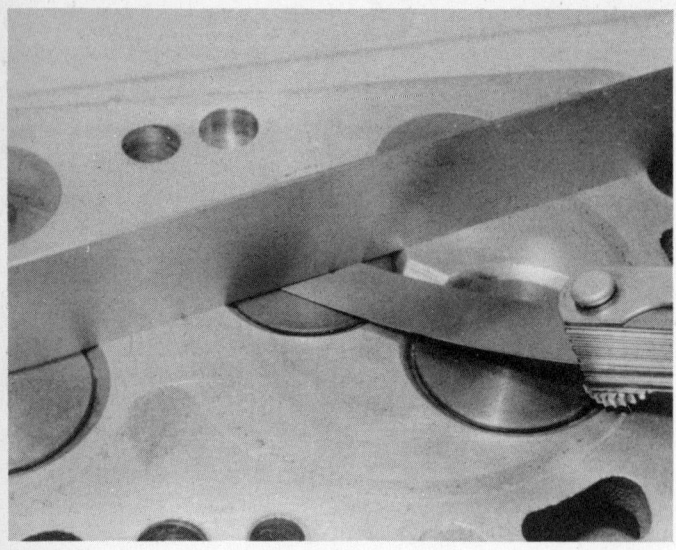

23.10 Checking a valve head recess

10 After refitting the valves, check the recess of their heads. This is a much coarser measurement and feeler blades will suffice (photo). The desired recess is given in the Specifications; it may be possible to correct out-of-limit valves by machining of the valve and/or seat. Seek specialist advice in this case.

11 Do not overlook the condition of core plugs, studs and dowels.

24 Cylinder head and pistons – decarbonising

1 Bearing in mind that the cylinder head is of light alloy construction and is easily damaged, use a blunt scraper or rotary wire brush to clean all traces of carbon deposits from the combustion spaces and the ports. The valve heads, steams and valve guides should also be freed from any carbon deposits. Wash the combustion spaces and ports down with paraffin and scrape the cylinder head surface free of any foreign matter with the side of a steel rule, or a similar article.

2 If the engine is installed in the car, clean the pistons and the top of the cylinder bores. If the pistons are still in the block, then it is essential that great care is taken to ensure that no carbon gets into the cylinder bores as this could scratch the cylinder walls or cause damage to the piston and rings. To ensure this does not happen, first turn the crankshaft so that two of the pistons are at the top of their bores. Stuff rag into the other two bores or seal them off with paper and masking tape. The waterways should also be covered with small pieces of masking tape to prevent particles of carbon entering the cooling system.

3 Press a little grease into the gap between the cylinder walls and the two pistons which are to be worked on. With a blunt scraper carefully scrape away the carbon from the piston crown, taking great care not to scratch the aluminium. Also scrape away the carbon from the surrounding lip of the cylinder wall. When all carbon has been removed, scrape away the grease which will now be contaminated with carbon particles, taking care not to press any into the bores. To assist prevention of carbon build-up the piston crown can be polished with a metal polish. Remove the rags or masking tape from the other two cylinders and turn the crankshaft so that the two pistons which were at the bottom are now at the top. Place rag or masking tape in the cylinders which have been decarbonised, and proceed as just described.

25 Examination and renovation – general

1 With the engine stripped and all parts thoroughly cleaned, every component should be examined for wear. The items listed in the Sections following should receive particular attention and where necessary be renewed or renovated.

2 So many measurements of engine components require accuracies down to tenths of a thousandth of an inch. It is advisable therefore to check your micrometer against a standard gauge occasionally to ensure that the instrument zero is set correctly.

3 If in doubt as to whether or not a particular component must be renewed, take into account not only the cost of the component, but the time and effort which will be required to renew it if it subsequently fails at an early date. (With this in mind, Astra/Kadett/Belmont owners may wish to renew the right-hand engine mounting as a precautionary measure.)

26 Engine components – examination and renovation

Crankshaft

1 Examine the crankpin and main journal surfaces for signs of scoring or scratches, and check the ovality and taper of the crankpins and main journals. If the bearing surface dimensions do not fall within the tolerance ranges given in the Specifications at the beginning of this Chapter, the crankpins and/or main journals will have to be reground.

2 In the absence of a micrometer, the crankshaft bearing running clearances can be determined by the use of Plastigage. This is a calibrated plastic filament which is placed between the bearing journal and shell. No oil must be present. The main bearing or connecting rod cap is fitted and its fastenings tightened to the specified torque, then the cap and shell are removed again. The width of the squashed filament is compared with the chart on the packet to determine the clearance (photo). This method is good for detecting taper, but it will not detect out-of-round unless several measurements are taken at different points.

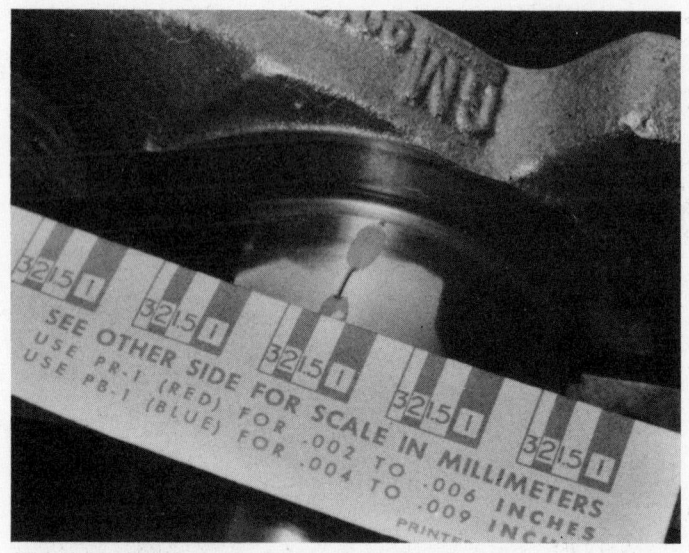

26.2 Using Plastigage to determine bearing clearance

3 Big-end and crankpin wear is accompanied by distinct metallic knocking, particularly noticeable when the engine is pulling from low revs, and some loss of oil pressure. Main bearing and main journal wear is accompanied by severe engine vibration and rumble – getting progressively worse as engine revs increase – and again by loss of oil pressure.

4 If the crankshaft requires regrinding take it to an engine reconditioning specialist, who will machine it for you and supply the correct undersize bearing shells. Note that the bearing journals must be hardened (by the process known as 'nitriding', or equivalent) after grinding. It is wise to check the clearances on the reground crankshaft, as described in paragraph 2, during reassembly.

Chapter 1 Engine

Big-end and main bearing shells

5 Inspect the big-end and main bearing shells for signs of general wear, scoring, pitting and scratches. The bearings should be matt grey in colour. With lead-indium bearings, should a trace of copper colour be noticed, the bearings are badly worn as the lead bearing material has worn away to expose the indium underlay. Renew the bearings if they are in this conditon or if there are any signs of scoring or pitting. **You are strongly advised to renew the bearings – regardless of their condition – at time of major overhaul. Refitting used bearings is a false economy.**

6 The undersizes available are designed to correspond with crankshaft regrind sizes. The bearings are in fact, slightly more than the stated undersize as running clearances have been allowed for during their manufacture.

7 Main and big-end bearing shells can be identified as to size by the marking on the back of the shell (photo). Typical markings are given in the Specifications; if other markings are encountered, consult a GM dealer for interpretation. Remember that undersize shells may have been fitted in production. As well as numbers, original equipment shells are colour coded on their edges.

Cylinder bores

8 The cylinder bores must be examined for taper, ovality, scoring and scratches. Start by carefully examining the top of the cylinder bores. If they are at all worn a very slight ridge will be found on the thrust side. This marks the top of the piston travel. The owner will have a good indication of the bore wear prior to dismantling the engine, or removing the cylinder head. Excessive oil consumption accompanied by blue smoke from the exhaust can be caused by worn cylinder bores and piston rings.

9 Measure the bore diameter across the block and just below any ridge. This can be done with an internal micrometer or a dial gauge. Compare this with the diameter of the bottom of the bore, which is not subject to wear. If no measuring instruments are available, use a piston from which the rings have been removed and measure the gap between it and the cylinder wall with a feeler gauge.

10 Refer to the Specifications. If the cylinder wear exceeds the permitted tolerances then the cylinders will need reboring. If the wear is marginal and within the tolerances given, proprietary piston rings which are claimed to offset the effects of bore wear can be fitted. The improvement made by such rings may be short-lived, however. Cylinders which have already been bored to the maximum oversize can have dry liners fitted to reclaim them – consult a GM dealer or other specialist.

Connecting rods

11 Examine the mating faces of the big-end caps to see if they have ever been filed in a mistaken attempt to take up wear. If so, the offending rods must be renewed.

12 Check the alignment of the rods visually, and if all is not well, take the rods to your local agent for checking on a special jig.

13 The weight of the connecting rod is indicated by a number or colour mark on the rod cap – see Specifications. Replacement rods should be of the same weight class.

14 Separation of the pistons from the rods is described in Section 11. Wear of the gudgeon pin or the small-end bush in the rod means that the components must be renewed; new gudgeon pins are not sold separately from new pistons.

Pistons and piston rings

15 If the pistons and/or rings are to be re-used, remove the rings from the pistons. Prepare three feeler blades or similar strips of thin metal, then spring the top ring open just far enough to insert the blades. The ring can then be slid off the piston without scratching or scoring (photo).

16 Repeat the process for the second and third rings.

17 Keep the rings with their original piston if they are to be re-used. The top ring carries a very small step on its upper face; the second ring has a 'TOP' marking which must be uppermost. The third (oil scraper) ring may be fitted either way up.

18 Inspect the pistons to ensure that they are suitable for re-use. Check for cracks, damage to the ring grooves and lands, and scores or signs of picking-up on the piston walls. If a micrometer is available, check the diameter of the piston around the base of the skirt.

19 Clean the ring grooves in the piston using a piece of old piston ring. Take care not to remove any metal or scratch the ring lands. Protect your fingers – piston rings are sharp!

20 Check the ring end gaps in the bores. Push the ring to the unworn lower section of the bore – use a piston to do this, so that the ring stays square. Measure the end gap with feeler blades and compare it with that specified. If the gap is too big (and the bore is unworn) the rings must be renewed.

21 It is worth checking the vertical clearance of the compression rings in their grooves. No limit is specified by the makers; experience

26.7 Identification markings on the back of a crankshaft bearing shell

26.15 Using three feeler blades for piston ring removal

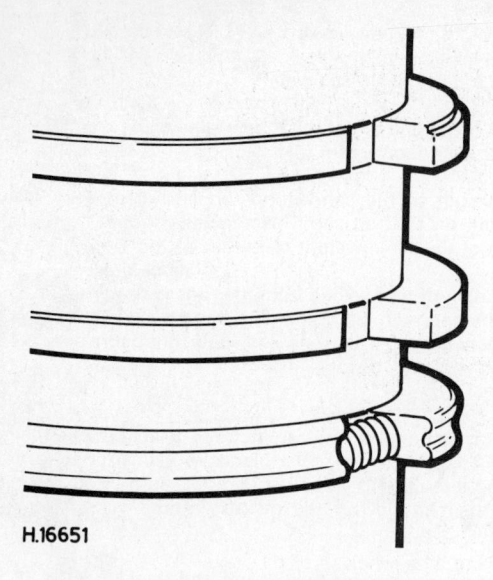

Fig. 1.18 Piston ring profiles (Sec 26)

suggests that a gap of up to 0.075 mm (0.003 in) is satisfactory, but much more than this may lead to oil pumping and rapid groove wear in service. New rings will only be effective here if the grooves are relatively unworn.

22 If buying new standard rings of GM manufacture, note that two grades are available, the smaller being for piston production grades 1 and 2 and the larger for grade 3.

23 Check the end gap of new rings at both top and bottom of the bores. If the gap is too small, the ends of the ring can be ground carefully until the correct gap is achieved.

24 If new pistons are to be fitted without reboring, they should be selected from the grades available (see Specifications) after accurate measurement of the bores. If a rebore is undertaken, it is normal for the repairer to obtain the correct oversize pistons at the same time. The piston size and the maker's identity are marked on the crown (photo).

25 Remember to deglaze old bores if new rings are fitted – see Section 11.

Flywheel

26 If the teeth on the flywheel starter ring are badly worn, or if some are missing, then it will be necessary to remove the ring and fit a new one.

27 Either split the ring with a cold chisel after making a cut with a hacksaw blade between two teeth, or use a soft-headed hammer (not steel) to knock the ring off, striking it evenly and alternately, at equally spaced points. Take great care not to damage the flywheel during this process and protect your eyes from flying fragments.

28 The new ring gear must be heated to 180° to 230°C (356° to 446°F). Unless facilities for heating by oven or flame are available, leave the fitting to your dealer or motor engineering works. The new ring gear must not be overheated during this work or the temper of the metal will be altered.

29 The ring should be tapped gently down onto its register and left to cool naturally when the contraction of the metal on cooling will ensure that it is a secure and permanent fit.

30 If the driven plate contact surface of the flywheel is scored or on close inspection shows evidence of small hair cracks, caused by overheating, it may be possible to have the flywheel surface ground provided the overall thickness of the flywheel is not reduced too much. Consult your specialist engine repairer and if it is not possible, renew the flywheel complete.

Fig. 1.19 Cleaning out a piston ring groove (Sec 26)

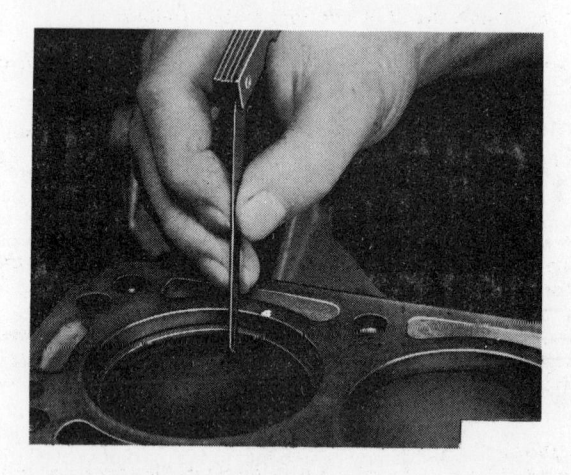

Fig. 1.20 Checking a piston ring end gap (Sec 26)

26.24 Maker's identity and bore grade markings on piston crown

Chapter 1 Engine

Driveplate (automatic transmission)
31 Should the starter ring gear on the driveplate require renewal, renew the driveplate complete.

Crankshaft spigot bearing
32 If the needle bearing in the centre of the crankshaft flange is worn, fill it with grease and tap in a close-fitting rod. Hydraulic pressure will remove it. Alternatively, a very small extractor having a claw type leg may be used. When tapping the new bearing into position, make sure that the chamfered side of the bearing enters first.

33 Later engines (from mid-1984) are not fitted with a spigot bearing.

Camshaft and carrier
34 With the camshaft removed, examine the bearings for signs of obvious wear and pitting. If evident, a new camshaft carrier will probably be required.

35 The camshaft itself should show no marks or scoring on the journal or cam lobe surfaces. If evident, renew the camshaft.

36 The camshaft thrust plate should appear unworn and without grooves. Renew it if not, or if the camshaft endfloat is excessive.

37 Always renew the camshaft front oil seal (photo).

38 Clean the filter in the camshaft cover in petrol or other solvent and allow it to dry.

39 If a new camshaft is being fitted, pay attention to the running-in schedule (Section 8).

Valve lifters, cam followers and thrust pads
40 The valve hydraulic lifters must be renewed if worn or if there is a history of noisy operation.

41 Inspect the cam followers and thrust pads for signs of wear or grooving; renew if evident. It is wise to renew the cam followers, if the camshaft is being renewed. Refer to Chapter 8 for further information.

Core plugs
42 Check the condition of the core plugs in the block and cylinder head. Those which close oily areas are unlikely to deteriorate, but those which close the water jacket may suffer from rust, especially if plain water has been used as coolant. Obviously it is much better to renew the plugs while they are accessible.

43 To remove a core plug, punch or drill a small hole in the middle of the plug. Screw in a large self-tapping screw and use a suitable lever on the screw head to extract the plug (photo).

44 Clean out the plug seat. Coat the sides of the new plug with sealant, then drive it into place.

27 Engine reassembly – general

1 All engine components must be clean before starting reassembly. The working area should also be clean and well-lit.

2 Obtain a set of new gaskets, oil seals etc. Also obtain new bolts where these are specified. Renew any other bolts, studs or nuts with damaged threads.

3 Besides the usual hand tools, a torque wrench will be required, as will some means of measuring angular rotation (eg a protractor and some card).

4 Have handy a squirt type oil can full of clean engine oil, plenty of clean rag, non-setting and RTV jointing compound and thread locking compound.

26.37 Fitting a camshaft front oil seal

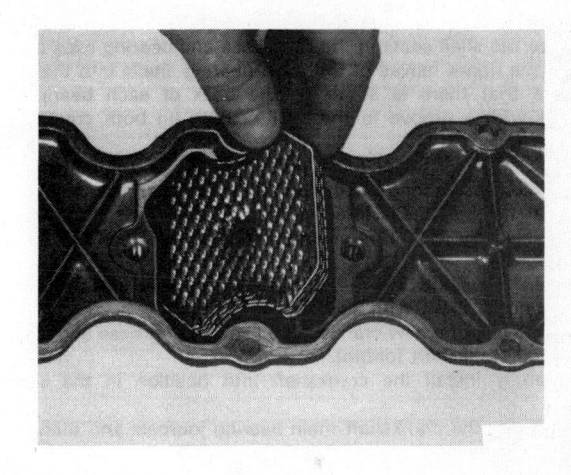

Fig. 1.21 Removing the filter from the camshaft cover (Sec 26)

26.43 Removing a core plug

Chapter 1 Engine

5 Refer to Section 6 and obtain one of the tools needed to set the valve timing. The dial test indicator is recommended since it will also be used for timing the fuel injection pump, and can be used for several other checks during engine rebuilding.

28 Engine – complete reassembly

Crankshaft and main bearings

1 Blow out the oilways in the crankcase and crankshaft with an air line, if available, then inject clean engine oil into them.

2 Clean the main bearing shell seats in the crankcase and the shells themselves. New shells may have a protective coating which should be carefully removed.

3 Fit the upper half main bearing shells to their seats. Make sure that the oil holes and the locating tangs are correctly positioned. All the shells are the same except for the centre one, which has thrust flanges (photo).

4 Apply oil to the bearing shells, then lower the crankshaft into position (photos). Turn it once or twice to make sure it is seated.

5 Clean the shell seats in the main bearing caps and fit the shells to the caps, again making sure that they are the right way round. As with the upper half shells, they are all the same except for the flanged centre shell.

6 Oil the shells and fit all the main bearing caps except the rear one (photo). Observe the identification marks or numbers on the caps, both to ensure correct numerical position and to ensure that the caps are the right way round. It may be necessary to strike the caps with a mallet to seat them – take care not to displace the shells if so. Do not tighten the cap bolts yet.

7 Coat the mating faces of the rear main bearing cap with non-setting jointing compound (to GM spec 15 04 200 or equivalent).

8 Lay a bead of RTV jointing compound (to GM spec 15 03 294, or equivalent) into the grooves on the sides of the rear main bearing cap. The bead should be approximately 6 mm (nearly a quarter inch) in diameter (photo).

28.3 Fitting the centre main bearing upper half shell – note thrust flanges

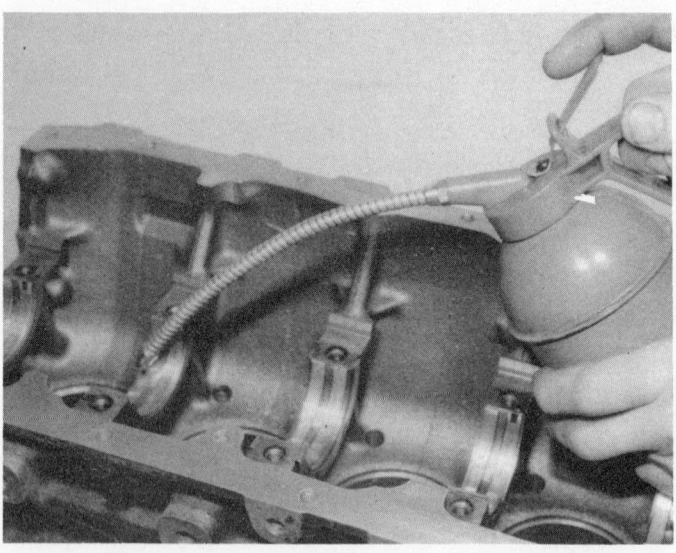

28.4A Lubricating the bearing shells

28.4B Lowering the crankshaft into position

28.6 Fitting the front main bearing cap

Chapter 1 Engine

Fig. 1.22 Applying jointing compound to the mating faces of the rear main bearing cap (Sec 28)

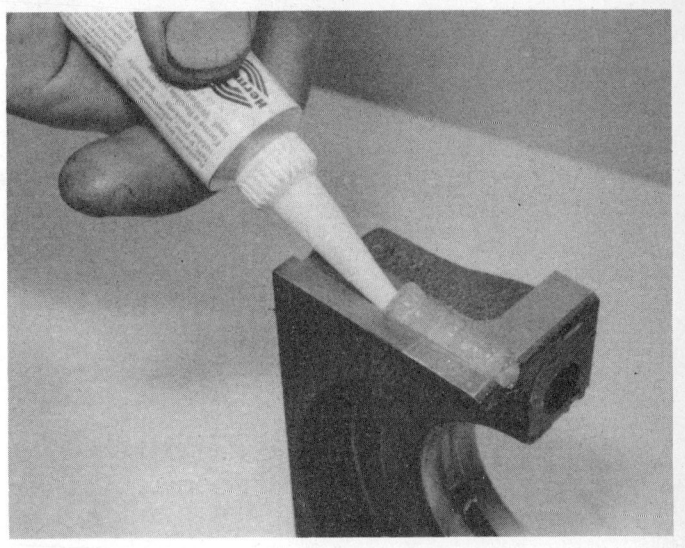

28.8 Applying sealant to the rear main bearing cap side grooves

9 Fit the rear main bearing cap and shell. When the cap is home, inject further RTV sealant into the grooves from the exposed end (photo).

10 Fit the main bearing cap bolts and nip them up. Make sure that the front main bearing cap is flush with the block face (photo).

11 Tighten the bearing cap bolts progressively to the specified torque.

12 Make sure that the crankshaft is still free to turn. Some stiffness is to be expected with new shells, but obvious tight spots or binding should be investigated. See Section 26, paragraph 2, for measurement of bearing clearances.

13 Measure the crankshaft endfloat, either with a dial test indicator on the flywheel flange, or using feeler blades at the centre bearing thrust flange (photo). Strike the shaft with a mallet at one end or the other to take up the endfloat. Out-of-limit endfloat must be due to wear, or to incorrect regrinding.

28.9 Injecting further sealant into the grooves from the exposed end

28.10 Front cap-to-block joint (arrows) must be flush

28.13 Measuring crankshaft endfloat

Chapter 1 Engine

Piston rings

14 The easiest method of fitting piston rings is to use feeler gauges (or similar) around the top of the piston and move the rings into position over the feelers. This is a reversal of the removal procedure detailed earlier in this Chapter.

15 Follow the manufacturer's instructions carefully when fitting rings to ensure that they are correctly fitted. Several variations of compression and oil control rings are available and it is of the utmost importance that they be located correctly in their grooves.

16 When the rings are in position check that the compression rings are free to expand and contract in their grooves. Certain types of multi-segment oil control rings are a light interference fit in their grooves and this may not therefore apply to them. When all the rings are in position on the pistons move them around to bring each ring gap to be some 180° away from the gap on the adjacent ring(s). When the oil control ring consists of two rails and a spacer, offset the upper rail gap 25 to 50 mm (1 to 2 in) to the left of the spacer gap; offset the lower rail gap a similar distance to the right.

Pistons and connecting rods

17 Reassemble the pistons and rods if they have been separated, making sure that they are the right way round. Insert the gudgeon pin circlips.

18 Lay out the pistons and rods in order, with the rod caps, bearing shells and cap nuts.

19 Clean the bearing shells and their seats. Press the shells into their positions, making sure that the locating tangs are the right way round.

20 Oil the piston rings and the cylinder bores. Make sure that the ring gaps are correctly spaced, then fit a piston ring compressor to one of the pistons.

21 Turn the crankshaft to bring the crankpin for the piston being fitted to BDC.

22 Offer the piston/rod assembly to the block, making sure that it is the right way round (arrow on piston crown pointing to pulley end) and that it is being fitted to the correct bore (photo).

23 Drive the piston into the bore by tapping the piston crown with, for example, the wooden handle of a hammer (photo). If resistance is encountered, release the compressor and see if a ring has jammed. Remove the compressor as the piston enters the bore.

24 Lubricate the crankpin and guide the connecting rod onto it, pushing the piston down the bore. Take care that the protruding studs do not scratch the crankpin, and check that the bearing shell is not displaced.

25 Fit the connecting rod cap, making sure that it is the right way round and that the shell is not displaced. Fit the securing nuts and tighten them to the specified torque (photos).

26 Repeat the operations to fit the other three pistons. Check after fitting each piston that the crankshaft can still be rotated.

Oil pump and sump

27 Lubricate the lips of the oil pump oil seal.

28 Fit a new oil pump gasket to the block, using a smear of grease to hold it in position.

29 Offer the oil pump to the block so that it fits onto the dowels (photo).

28.22 Offering the piston to the block – ring compressor fitted

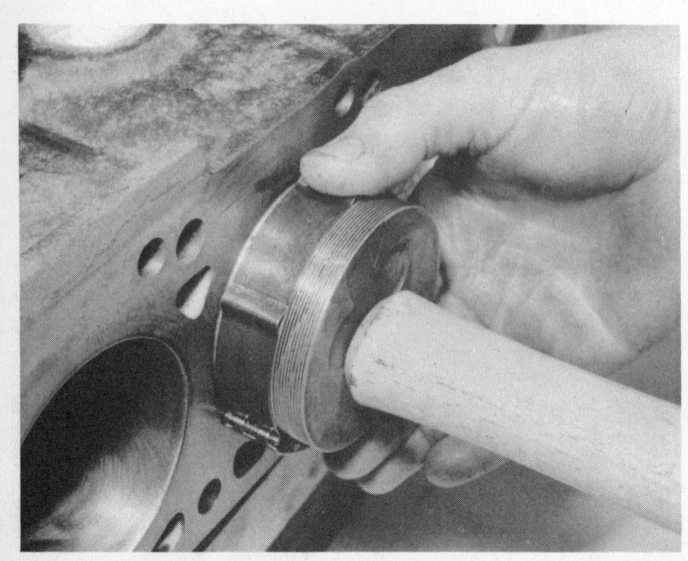

28.23 Driving the piston into the bore

28.25A Fitting a connecting rod cap

Chapter 1 Engine

28.25B Tightening a connecting rod nut

28.29 Fitting the oil pump – note new gasket on block

30 Fit and tighten the pump securing screws (photo).

31 Fit the oil pump pick-up tube, using a new O-ring between the tube flange and the pump. Secure the tube with two bolts at the flange and a single bolt at the block (photos).

32 Apply RTV jointing compound (to GM spec 15 03 294) to the joints between the rear main bearing cap and the block, and the oil pump and the block.

33 Smear a new sump gasket with grease and positon it on the block. Fit the sump, taking care not to displace the gasket (photo).

34 Fit the sump retaining bolts and washers, using thread locking compound on the bolt threads. Tighten the bolts evenly to the specified torque (photo).

Rear oil seal and flywheel
35 Lubricate the lips of a new rear oil seal and fit it to the crankshaft, using a plastic collar as described in Section 13. (There is a temptation to fit this seal to the crankshaft before fitting the main bearing caps, but there is then a risk of distorting or pinching the seal as the rear main bearing cap is tightened).

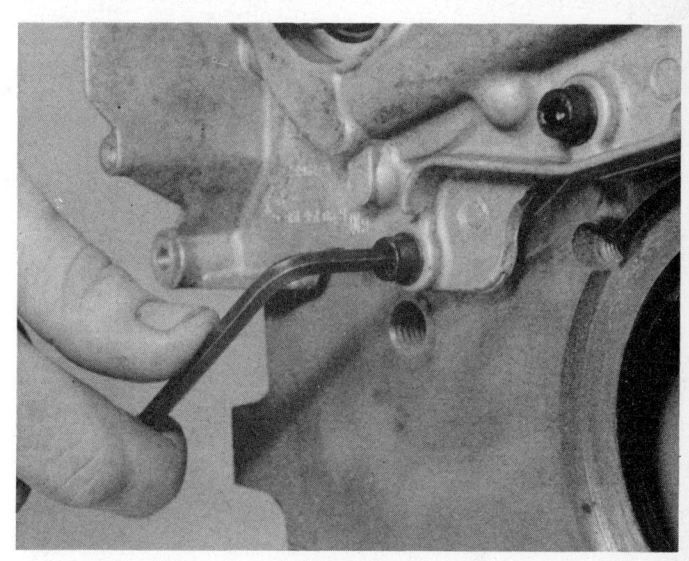

28.30 Tightening an oil pump securing screw

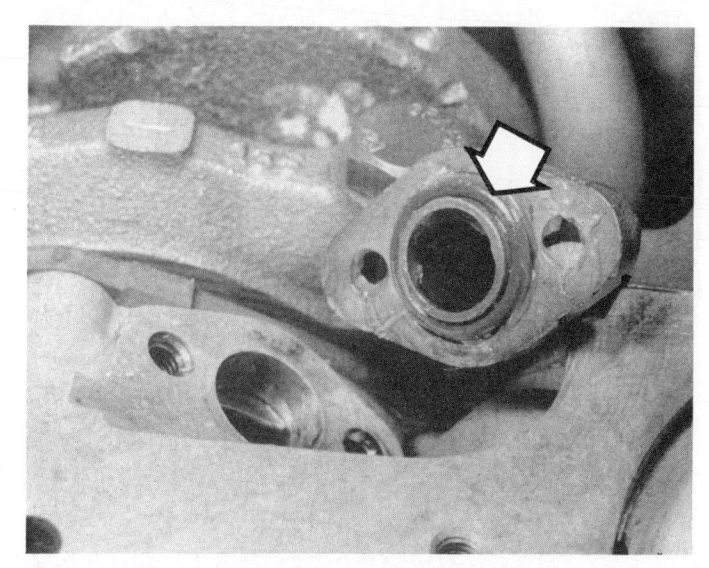

28.31A Fitting the oil pump pick-up tube – O-ring is arrowed

28.31B Oil pump pick-up tube flange bolts (arrowed)

28.31C Oil pump pick-up tube securing bolt (arrowed) on block

28.33 Fitting the sump

28.34 Tightening a sump bolt

36 Fit the flywheel (or driveplate) to the crankshaft flange, lining up the bolt holes for the shouldered bolt.

37 Apply thread locking compound to the threads of the new flywheel bolts. Fit the bolts, making sure the shouldered bolt is correctly located, and tighten them evenly to the Stage 1 specified torque.

38 Arrange some means of measuring the angular rotation of the bolts, then tighten them through the angle specified for Stage 2. It will be necessary to lock the flywheel as this is done: an old clutch cover bolt can be inserted from the starter motor side, but it is sure to be bent under the strain (photos).

Head gasket selection
39 As mentioned elsewhere, there are three thicknesses of head gasket. Selection of the correct gasket is determined by measuring piston protrusion above the block face at TDC. Precise figures are given in the Specifications.

40 The projection could be measured with a straight-edge and feeler blades, but it is more accurate (and much easier to find TDC) to use a dial test indicator.

28.38A Tightening a flywheel bolt through a specified angle

28.38B Scrap bolt used to lock flywheel

Chapter 1 Engine

41 Invert the engine so that it is standing on its sump. Chock it with wooden blocks if necessary to ensure stability.

42 Rotate the crankshaft to bring two pistons to TDC. Take care to determine, either with the dial test indicator or by eye, that the pistons are truly at the very top of their stroke.

43 If a dial test indicator is being used, zero it on the block face next to the piston, then measure the height of the piston crown above the block face (photo). If a straight-edge is being used, rest it on the two piston crowns and determine the thickness of feeler blades(s) which will just pass between the straight-edge and the block.

44 Bring the other two pistons to TDC and repeat the measurement, then select the head gasket.

Cylinder head and valve gear

45 Make sure that the block and head mating faces are clean and that the locating dowels are in positon in the block. Bring the pistons to mid-stroke (90° BTDC).

46 Fit the head gasket to the block with the OBEN/TOP mark uppermost. It will only fit one way. *Do not use any grease or jointing compound.*

47 Lower the assembled cylinder head onto the block, making sure that the dowels engage (photo).

48 Lubricate and fit the valve lifters, thrust pads and cam followers. If new valve lifters are being used, initially immerse each one in a container of clean engine oil and compress it (by hand) several times to charge it.

49 Apply a little non-setting jointing compound to the mating faces, then fit the assembled cam carrier and camshaft (photo).

50 Insert the new cylinder head bolts, not forgetting the washers. Nip up the bolts in spiral sequence from the centre outwards, being careful to pull down the cam carrier evenly.

51 In the same sequence tighten the bolts to the Stage 1 specified torque, then through the angles specified for Stages 2, 3 and 4. It becomes increasingly difficult to hold the engine still if it is on the bench, so if preferred the later tightening stages can be left until the engine is back in the vehicle. Attach a label or some other reminder where it will be seen before the engine is started.

Camshaft drivebelt and associated components

52 Fit the thermostat housing and thermostat, using new seals (photos).

53 Refit the Woodruff key to the crankshaft nose. Fit the crankshaft sprocket and its securing bolt. Lock the flywheel and tighten the sprocket bolt to the specified torque (photos).

54 Refit the water pump, using a new seal. Insert but do not tighten its securing bolts.

55 Fit the drivebelt backplate, then fit and secure the idler pulley (photo).

56 Refit the fuel pump mounting bracket (photo), then refit the fuel pump as described in Chapter 3, Section 8. Fit the Woodruff key and the pump sprockets. Tighten the sprocket centre nut.

57 Fit the camshaft sprocket and a new retaining bolt, but do not tighten the bolt yet (photo).

58 Bring the crankshaft to TDC, No 1 firing. Similarly align the pump sprocket timing mark and bring the camshaft to approximately the correct position, ie both lobes for No 1 cylinder inclined upwards and those for No 4 downwards. Proceed carefully to avoid piston/valve contact.

28.43 Using a dial test indicator to measure piston projection

28.47 Fitting the cylinder head

28.49 Fitting the camshaft carrier

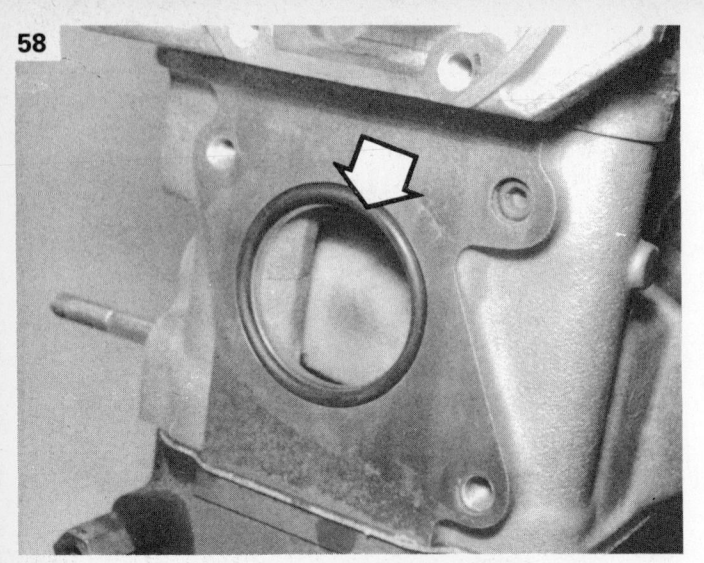

28.52A Fit a new O-ring (arrowed) to the thermostat housing recess

28.52B Tightening a thermostat housing bolt

28.53A Fitting the crankshaft sprocket

28.53B Tightening the crankshaft sprocket bolt

28.55 Securing the drivebelt idler pulley

28.56 Fitting the fuel injection pump bracket

Chapter 1 Engine

28.57 Camshaft sprocket loosely fitted

28.63 Refitting the dipstick tube

59 Fit the camshaft drivebelt around the sprockets and the idler pulley. Observe the running direction of the old belt if it is being re-used.

60 Adjust the drivebelt tension as described in Section 5.

61 Adjust the valve timing as described in Section 6.

62 Adjust the injection pump timing if necessary as described in Chapter 3, Section 7. Refit and secure the injection pipes.

63 Refit the dipstick tube, using a new gasket. Note that a cable bracket is secured by the same bolts (photo).

64 Refit the vacuum pump, using new seals.

65 Refit the large coolant pipe, using new seals (photo).

66 Refit the cam cover, using a new gasket.

67 Refit the breather pipe between the dipstick tube and the cam carrier (photo).

68 Refit the camshaft drivebelt covers. Remember that one of the cover bolts also secures a fuel pipe clip.

69 Refit the clutch, when applicable.

70 Refit and secure the manifolds, using new gaskets.

71 The oil filter and crankshaft pulley are best left until after the engine has been refitted. However, it is recommended that the alternator be refitted (with its engine mounting bracket, when applicable) while the engine is still out and access is good.

29 Engine and transmission – reconnection

Manual transmission

1 If the clutch has been disturbed, make sure that the driven plate is centralised.

2 Offer the transmission to the engine, twisting it back and forth slightly so that the splines on the input shaft enter the clutch driven plate. Do not allow the weight of the transmission to hang on the input shaft.

3 Engage the transmission with the dowels on the engine. Fit and tighten the engine-to-transmission bolts.

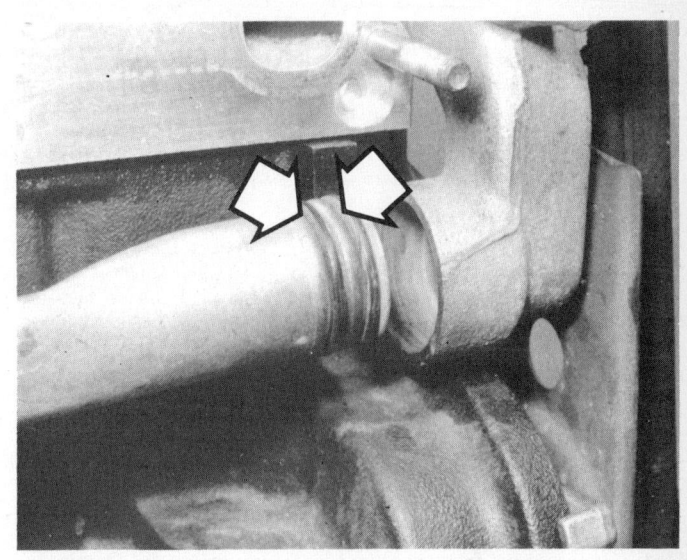

28.65 Use new seals (arrowed) on the large coolant pipe

28.67 Fitting the breather pipe between the dipstick tube and the cam carrier

Chapter 1 Engine

4 Refit the starter motor and its heat shield. Fit the motor with its leads attached, since there is no access with the heat shield fitted.

5 Refit the clutch/flywheel access plate.

Automatic transmission

6 Make sure that the torque converter is fully engaged with the transmission (see Fig. 1.23).

7 Offer the transmission to the engine so that the coloured spot on the torque converter is as close as possible to the white spot on the driveplate.

8 Engage the transmission with the dowels on the engine, then fit and tighten the engine-to-transmission bolts.

9 Secure the torque converter to the driveplate using new bolts. Turn the crankshaft to bring the bolt holes into reach through the cover plate aperture.

10 Refit the starter motor and heat shield – see paragraph 4.

11 Refit the torque converter/driveplate access plate.

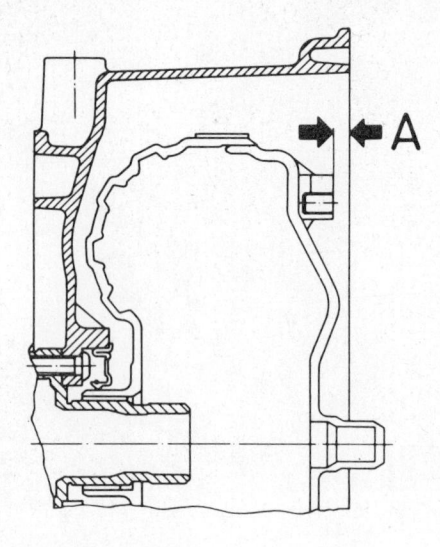

30 Engine and transmission – refitting

1 With the front of the vehicle raised and securely supported, position the engine/transmission unit under the engine bay. Connect the lifting tackle.

2 Carefully raise the unit into the engine bay. One person should operate the lifting tackle while another guides the unit into place and keeps watch for fouling.

3 Fit the mounting brackets to the engine/transmission and bodyframe. Some lifting and lowering of the unit will be necessary before all the bolts will go in. Do not tighten the bolts fully until all the mountings are fitted.

4 Remove the engine lifting tackle.

5 Insert the driveshafts into the transmission, being careful not to damage the oil seals. Make sure that the retaining snap-rings engage inside the transmission (photo).

6 Reconnect and secure the front suspension bottom balljoints. Do not forget the split pins or spring clips which secure the balljoint nuts (photo).

Fig. 1.23 Torque converter fully engaged (Sec 29)

A = 9 to 10 mm (0.35 to 0.39 in)

7 Refit the crankshaft pulley, if it was left off during engine refitting, and tighten its screws (photo).

8 Fit a new oil filter element – see Section 4.

9 Refit the alternator (if not already done) and tension its drivebelt.

10 Refit the power steering pump, when applicable, and tension its drivebelt.

11 Refit the throttle and cold start cables to the fuel inejction pump. Refer to Chapter 3, Sections 13 and 14.

12 Reconnect the fuel supply and return hoses to the fuel injection pump. Remember to secure the hoses to the camshaft drivebelt cover (photo).

13 Refit the exhaust downpipe, using a new flange gasket (photo) and (if necessary) a new sealing ring at the flexible joint. Secure the

30.5 Left-hand driveshaft ready for inserting into transmission

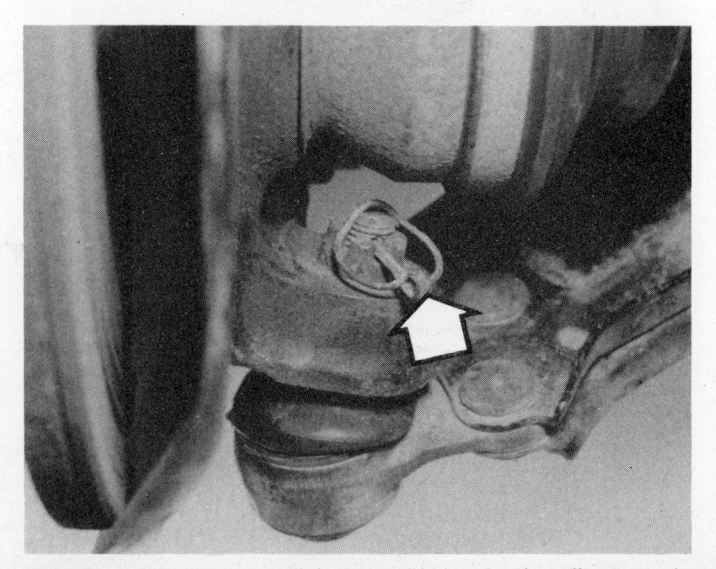

30.6 Suspension bottom balljoint assembled – securing clip arrowed

Chapter 1 Engine 61

30.7 Crankshaft pulley-to-sprocket screws (arrowed)

30.12 Fuel hose clip secured by drivebelt cover screw

30.13 New gasket on the exhaust downpipe flange

pipe-to-transmission bracket, when fitted. In other versions the downpipe may be secured to an engine/transmission mounting.

14 Reconnect the gearshift linkage or cable. Refer to the appropriate manual for petrol-engined vehicles for adjustment procedures.

15 Reconnect and adjust the clutch cable or kickdown cable.

16 Reconnect the speedometer cable. Secure it as far as possible away from the exhaust manifold and downpipe, without straining the cable.

17 Refit the radiator and cooling fan, if they were removed.

18 Reconnect the cooling system hoses. Also reconnect the automatic transmission cooler lines, when applicable.

19 Refill the cooling system.

20 Reconnect the engine electrical services. Also reconnect the starter motor lead to the battery and the starter motor command lead.

21 Refit the air cleaner and its snorkel.

22 Refill the engine with oil.

23 Check the transmission oil level and top up if necessary.

24 Reconnect the transmission earth strap.

25 Reconnect the battery earth lead.

26 Run the engine, referring first to Section 31, then refit the bonnet if it was removed.

27 Refit the front wheels, if not already done.

31 Engine – initial start-up after overhaul

1 If the final stages of head bolt tightening were deferred when the engine was on the bench, carry them out now (up to and including Stage 4).

2 Make sure that there is sufficient clean fuel in the tank, that the battery is well charged, and that all lubricants and fluids have been replenished.

3 A good deal of cranking on the starter motor may be required to prime the fuel system. (It is self-bleeding.) Do not operate the starter for more than 10 seconds at a time, then pause for a similar period to allow the battery and starter motor to recover.

4 As soon as the engine is running, check that the oil pressure light has gone out. This will take a few seconds as the oil filter fills up. There may also be some noise from the valve gear until the hydraulic lifters are properly pressurized with oil.

5 Check for leaks of oil, coolant and fuel. Be prepared for some smoke and fumes as assembly lubricant burns off.

6 Allow the engine to reach operating temperature, then adjust the idle speed as described in Chapter 3.

7 Stop the engine and check the tightness of the exhaust downpipe joints. Also check for oil or coolant leaks while stopped.

8 Carry out the Stage 5 tightening of the cylinder head bolts.

9 Check engine oil and coolant levels and top up if necessary. Take precautions against scalding if the engine is still hot.

10 Road test the vehicle to check that the engine is performing acceptably.

11 If new bearings and/or pistons have been fitted, the engine should be run in at reduced speed and load for the first 600 miles (1000 km) or so. Change the engine oil and filter after this mileage.

12 Also after the first 600 miles (1000 km), the cylinder head bolts should be tightened through the angle specified for Stage 6. This should be done cold or warm (not hot).

32 Fault diagnosis – engine

Symptom	Reason(s)
Engine does not turn when starter operated	Battery discharged or defective Battery connections loose or corroded Transmission earth strap loose or broken Starter switch or wiring defective Starter motor or solenoid defective Automatic transmission not in P or N Automatic transmission starter inhibitor switch defective
Engine turns but will not start	Insufficient cranking speed (battery or starter defective) Incorrect starting procedure (see operator's handbook) Poor compression (see below) Valve timing incorrect Fuel system fault (see Chapter 3) Exhaust system blocked
Engine stalls and will not restart	Fuel tank empty Overheating Other fuel system fault (see Chapter 3) Camshaft drivebelt broken or slipped Major mechanical failure
Engine misfires or idles unevenly	Fuel system fault (see Chapter 3) Valve timing incorrect Valve(s) burnt or sticking Valve spring(s) weak or broken Camshaft badly worn
Poor compression	Valve(s) burnt or sticking Valve spring(s) weak or broken Worn pistons, rings or bores Blown head gasket Valve timing incorrect
Lack of power	Air cleaner clogged Exhaust system restricted Other fuel system fault (see Chapter 3) Poor compression (see above) Brakes binding
Excessive oil consumption	External leakage Valve stem oil seals worn Valve stems or guides worn Worn pistons, rings or bores Crankcase ventilation hoses clogged or wrongly routed Inferior or incorrect grade of oil
Unusual noises	Valve gear worn * Fuel system defect (see Chapter 3) Camshaft drivebelt worn or incorrectly tensioned Worn pistons, rings or bores Crankshaft bearings worn Peripheral component fault (water pump, alternator, vacuum pump etc)

*It is normal for a considerable amount of noise to come from hydraulic valve lifters on initial start-up after overhaul. This should only continue until the valve lifters are properly pressurized with oil. Additionally, on a high-mileage engine, there may be some initial noise if the engine has not been started for a period of time.

Chapter 2 Cooling system

For modifications, and information applicable to later models, see Supplement at end of manual

Contents

Antifreeze mixture – general	11	Fault diagnosis – cooling system	12
Cooling system – draining	3	General description	1
Cooling system – filling	5	Maintenance and inspection	2
Cooling system – flushing	4	Radiator – removal, inspection and refitting	6
Fan – testing, removal and refitting	8	Thermostat – removal, testing and refitting	7
Fan thermoswitch – testing, removal and refitting	9	Water pump – removal and refitting	10

Specifications

General
System type .... Water-based coolant, sealed pressurised system, thermostatically controlled
Radiator cooling .... Ram air and electric fan

Thermostat
Opening commences at .... 91°C (196°F)
Fully open at .... 106°C (223°F)
Marking 91

Fan thermoswitch
Cuts in at .... 97°C (207°F)
Cuts out at .... 93°C (199°F)

Radiator cap
Blow-off pressure .... 1.20 to 1.35 bar (17.4 to 19.6 lbf/in²)
Marking 120

Coolant
Type/specification .... Ethylene glycol based antifreeze to GME 13368 (Duckhams Universal Antifreeze and Summer Coolant)
Capacity .... 7.7 litre (13.6 pint) approx

Water pump
Type .... Impeller
Drive .... From camshaft drivebelt

Torque wrench settings

	Nm	lbf ft
Water pump to block	25	18
Thermostat housing to head	15	11
Temperature sensor to thermostat housing	8	6

Chapter 2 Cooling system

1 General description

The engine is water-cooled. The cooling system is sealed and pressurised, in line with current practice. Waste heat from the cooling system is used in the vehicle heater.

Coolant (commonly referred to as water) circulates round the engine water jacket and passages under the influence of a pump driven by the camshaft drivebelt. When the engine is cold, a thermostat prevents the coolant from leaving the engine except to feed the heater. As operating temperature is reached, the thermostat opens and coolant is allowed to flow through the radiator.

The radiator acts as a heat exchanger between the coolant and the outside air. In normal driving the flow of air past the radiator provides sufficient cooling effect. In heavy traffic or similar conditions, a thermostatic switch in the radiator operates an electric cooling fan to boost the airflow.

Expansion and contraction of the coolant is accommodated by an expansion tank, which incorporates level markers and a filler cap. The filler cap has vacuum and pressure relief valves. The pressure relief valve does not operate until a pressure considerably above atmospheric is reached; the effect of this is to raise the boiling point of the coolant and so increase the efficiency of the system.

The precise location of components such as the expansion tank, and consequently the routing of the hoses, may vary between models.

2 Maintenance and inspection

1 Check the coolant level frequently. With the system cold, the level should be up to the KALT (cold) line on the expansion tank; with the system hot the level should be somewhat higher. The expansion tank is translucent so there is no need to remove the cap to check the level.

2 If topping-up is necessary, *take great care if the system is hot*. Unscrew the filler cap slowly, a little at a time, and wait for any pressure release to stop before removing the cap (photo). Severe scalding may result from incautious removal.

3 With the cap removed, top up if necessary with the correct antifreeze mixture (see Section 11). Check for leaks if frequent topping-up is required. Coolant which is not being lost through external leaks must be getting into the engine via the head gasket or cracks in the head or block. A pressure test of the cooling system can be performed by most garages and can help to locate leaks.

4 At every major service interval, inspect the coolant hoses and their clips for security and good condition. Pay particular attention to the heater hoses where they pass near the exhaust downpipe, and to the spur to the expansion tank. When routed as shown in Fig. 2.2, this spur can suffer from chafing; protect it if necessary by lagging it with suitable adhesive tape. Renew any hoses or clips in poor condition after draining the system as described in Section 3.

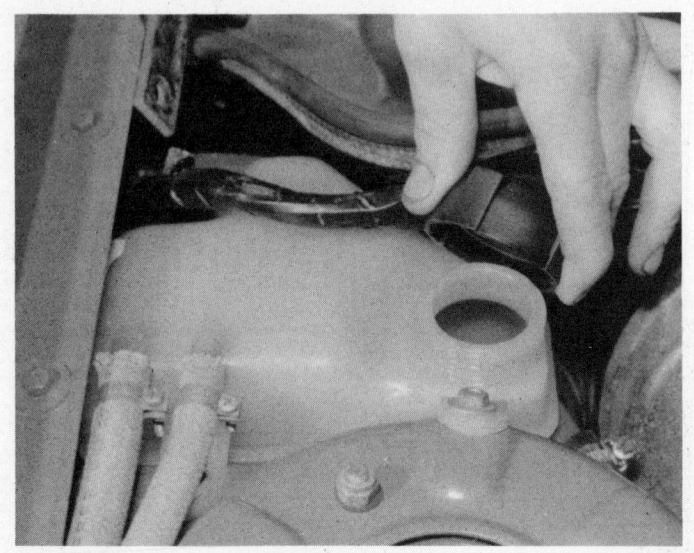

2.2 Removing the expansion tank filler cap

Fig. 2.1 Check that heater hoses are secured clear of the exhaust downpipe by plastic ties (arrowed) (Sec 2)

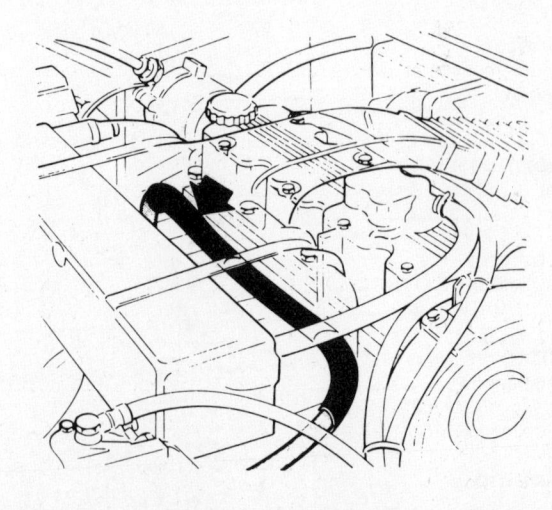

Fig. 2.2 Hose to expansion tank can be protected by lagging (arrowed) (Sec 2)

Chapter 2 Cooling system

2.5 Using an antifreeze tester. Protection against freezing is shown by the position of the balls

3.3 The radiator bottom hose

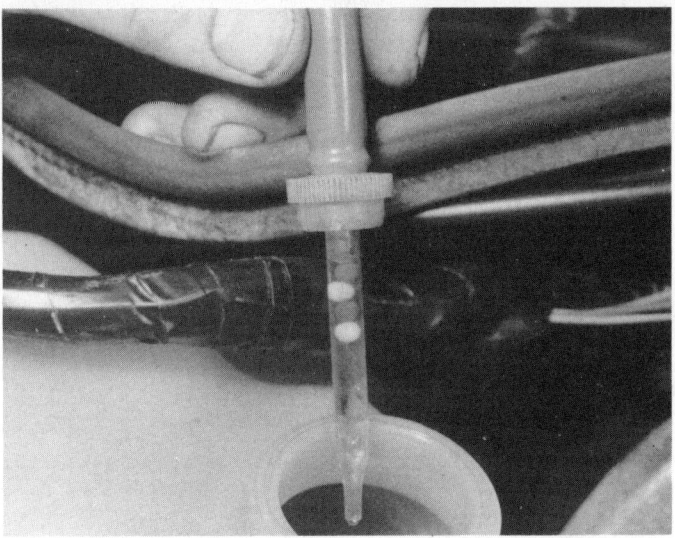

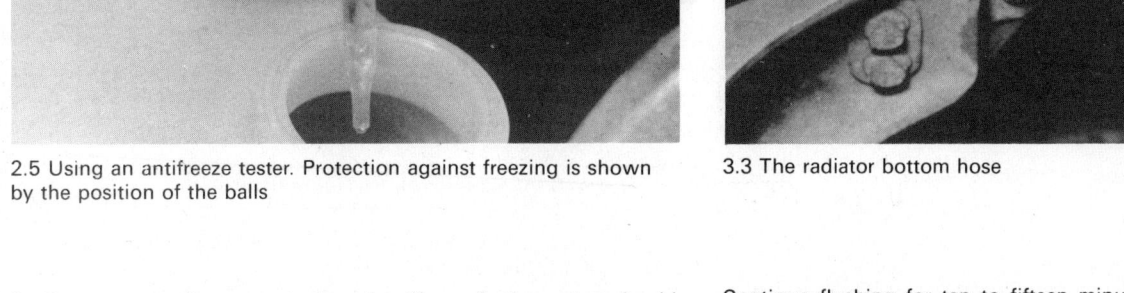

5 Every autumn the concentration of antifreeze in the system should be checked and made good if necessary. Most garages can do this check, or an instrument similar to a battery hydrometer can be purchased for making the check at home (photo).

6 It is recommended that the coolant be renewed every two years. At the same time the opportunity should be taken to flush the system, if necessary, and to renew any hoses or clips in poor condition. Regular renewal of the coolant will maintain its anti-corrosive properties and help to prevent sludge and scale build-up. Refer to Section 11.

3 Cooling system – draining

1 The system should only be drained when it is cold. If it must be drained hot, take great care to avoid scalding.

2 Remove the expansion tank cap.

3 Place a container underneath the radiator bottom hose. Disconnect the hose from the radiator and allow the system to drain (photo).

4 There is no cylinder block drain plug, so it is impossible to drain the system completely.

4 Cooling system – flushing

1 If coolant has been neglected, or if the antifreeze mixture has become diluted, then in time the cooling system will gradually lose efficiency as the cooling passages become choked with rust, scale deposits and other sediment. To restore cooling system efficiency it is necessary to flush the system clean.

2 First drain the system, as described in the previous Section, and then remove the thermostat, as described in Section 7. Temporarily refit the thermostat housing and reconnect the hose.

3 Insert a garden hose into the disconnected radiator bottom hose and secure it in place with rags. Turn on the supply and allow clean water to flow through the system and out of the radiator bottom outlet.

Continue flushing for ten to fifteen minutes or until clean rust-free water emerges from the radiator.

4 If the contamination is particularly bad, reverse flush the system by inserting the garden hose in the radiator bottom outlet and allow the water to flow through the system and out of the radiator bottom hose. This should dislodge deposits that were not moved by conventional flushing. If any doubt exists about the cleanliness of the radiator after flushing, it should be removed, as described in Section 6, so that it can be flushed and agitated at the same time. After severe flushing carry out a normal flow flush before refitting the thermostat and reconnecting the system hoses.

5 In extreme cases the use of a proprietary de-scaling compound may be necessary. If such a compound is used, adhere to the manufacturer's instructions and satisfy yourself that no damage will be caused to the engine or cooling system components.

6 If the coolant is renewed regularly, flushing will not normally be required, simply drain the old coolant and refill with a fresh mixture.

5 Cooling system – filling

1 Make sure that all hoses and clips are in good condition. Refit any disturbed hoses and see that their clips are tight.

2 Fill the system via the expansion tank cap. If new coolant is being put in, start by pouring in the required quantity of neat antifreeze and follow it up with the water.

3 Massage the large coolant hoses as filling proceeds to help displace air pockets.

4 Most vehicles will be fitted with a self-venting cooling system – this can be recognised by the two small vent hoses which enter the top of the expansion tank. If the system is not self-venting, open the bleed screw on the thermostat elbow during filling and close it when coolant runs out at the bleed screw (photo).

5 When the system appears full, refit the expansion tank cap. Run the engine up to operating temperature, keeping a look-out for coolant leaks, then stop it and allow it to cool. Recheck the coolant level and top up if necessary.

6 Recheck the tightness of all hose clips when the engine has cooled, and again after a few hundred miles.

Chapter 2 Cooling system

5.4 Bleed screw (arrowed) on thermostat elbow

6 Radiator – removal, inspection and refitting

The radiator can be removed with the fan still attached, or the fan may be removed first as described in Section 8.

1 Drain the cooling system as described in Section 3. Save the coolant if it is fit for re-use.

2 Disconnect the top hose and (if applicable) the vent hose from the radiator. On automatic transmission models, disconnect and plug the fluid cooler pipes; be prepared for fluid spillage.

3 Disconnect the fan and thermoswitch wiring, if not already done. Also remove the air cleaner snorkel.

4 Release the radiator from its mountings. These will vary with model and year; typically there are two bolts or clips at the top, and a couple of rubber blocks to receive pegs at the bottom (photo).

5 Carefully lift out the radiator, with fan if not already removed. Remember that the matrix is easily damaged.

6 Clean dead insects, leaves and other debris from the radiator matrix using compressed air or water and a soft brush. Inspect the matrix for leaks. Proprietary compounds are available which may be effective in stopping leaks in the short term, but for permanent repair a GM dealer or radiator specialist should be consulted.

7 Flush the radiator by running clean water through it, shaking it gently at the same time.

8 Refit the radiator in the reverse order of removal. Make sure that the mounting components and hoses are in good condition.

9 Refit the fan, if it was removed separately.

10 Refill the cooling system as described in Section 5.

11 On automatic transmission models, reconnect the fluid cooler pipes, then check the transmission fluid level and top up if necessary.

7 Thermostat – removal, testing and refitting

1 Drain the cooling system as described in Section 3. Save the coolant if it is fit for re-use.

2 Unbolt the thermostat elbow from the thermostat housing. Remove the elbow and lift out the thermostat (photo).

3 Test the thermostat by suspending it in a saucepan of cold water and bringing the water to the boil. The thermostat should not contact the saucepan. The thermostat should be closed until the water is nearly boiling, when it should start to open. The precise opening temperature is given in the Specifications. It is not so easy to check the fully open temperature, since this is above the boiling point of water in an open pan.

4 Renew the thermostat if it does not open and close as described, or if it is obviously damaged.

5 Refit the thermostat in the reverse order of removal. Always use a new sealing ring (photo).

6 Refill the cooling system as described in Section 5.

6.4 Removing a radiator securing clip

7.2 Removing the thermostat from the housing

Chapter 2 Cooling system

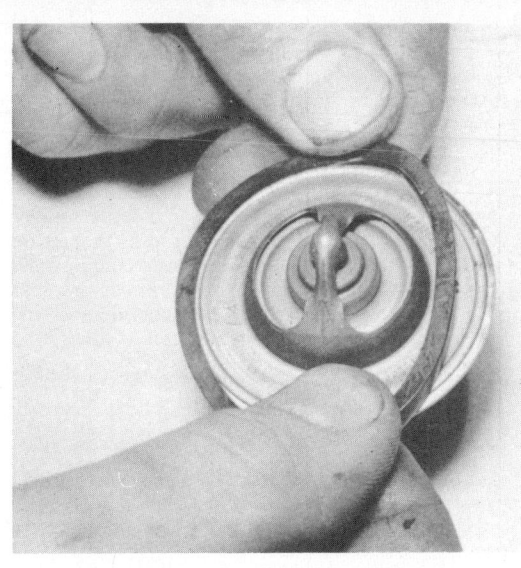

7.5 Fitting a new sealing ring to the thermostat

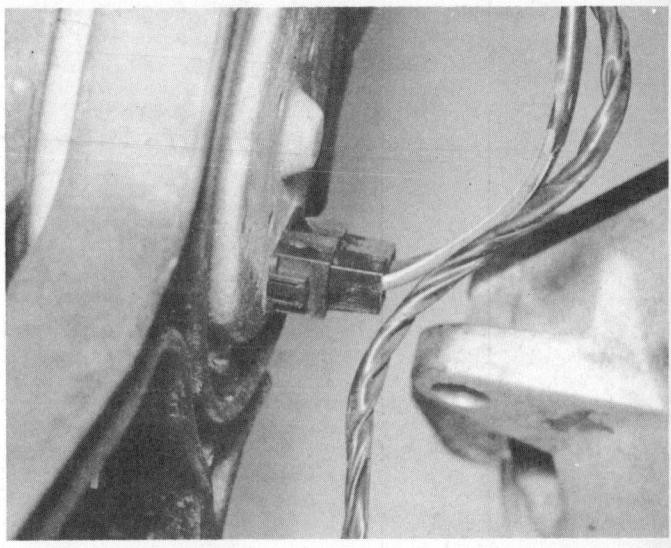

8.2 Fan wiring connector

8 Fan – testing, removal and refitting

1 If the fan does not operate even though the coolant is boiling, disconnect the leads from the thermoswitch (at the side of the radiator) and join them together with a paper clip or wire link. Do not allow the link to touch other metal. With the ignition on and the switch leads joined, the fan should run continuously. This test determines, in cases of fan malfunction, whether it is the fan or the thermoswitch which is at fault. See Section 9.

2 If the fan will not run as just described, unplug its wiring connector (photo) and use a test lamp or meter to check for voltage across the connector terminals. If no voltage is present, check the fuse and wiring. If voltage is present but the fan does not turn, it is defective and should be removed.

3 To remove the fan, unplug its wiring connector, then remove the two screws which secure the top corners of its cowl. Disengage the bottom corners from their mountings and carefully remove the fan and cowl (photo).

4 The fan can be removed from the cowl after undoing the three retaining nuts (photo).

5 It is not recommended that dismantling of the motor be attempted; spares are not available in any case.

6 Refit the fan to the cowl and secure it with the three nuts.

7 Refit and secure the fan and cowl to the radiator. Plug in the wiring connector.

9 Fan thermoswitch – testing, removal and refitting

1 If the fan runs all the time with the ignition on, or does not run even though the coolant is boiling, the thermoswitch may be at fault. Refer to paragraph 1 of Section 8.

2 Before condemning the thermoswitch, be sure that the coolant is really hot enough to trip the switch and that the problem is not (for instance) a faulty temperature gauge. A defective switch must be renewed.

3 To remove the switch, drain the cooling system as described in Section 3, saving the coolant if it is fit for re-use.

8.3 Removing the fan and cowl

8.4 Fan-to-cowl retaining nuts (arrowed)

4 Unplug the switch and unscrew it from its location in the radiator.

5 Screw the new switch in, using a little sealant on the threads, and plug in its connector.

6 Refill the cooling system as described in Section 5.

10 Water pump – removal and refitting

1 Drain the cooling system as described in Section 3.

2 Remove the camshaft drivebelt as described in Chapter 1, Section 5.

3 Remove the drivebelt idler pulley and backplate.

4 Remove the three bolts which retain the water pump. Lift out the pump: a good quantity of coolant will be released. It may be necessary to remove the alternator to provide sufficient clearance to remove the pump completely.

5 Pump overhaul is not a DIY job, since press facilities are required. Consult a GM dealer or engineering shop.

6 Use a new O-ring when refitting the pump. In order to prevent corrosion and the resultant impossibility of moving the coolant pump to tension the timing belt, apply silicone grease to the pump O-ring and cylinder block mating surface. Insert but do not tighten the retaining bolts (photos).

7 Refit the drivebelt backplate and the idler pulley.

8 Refit and tension the camshaft drivebelt as described in Chapter 1, Section 5.

9 Refit the other disturbed components, then refill the cooling system as described in Section 5.

11 Antifreeze mixture – general

1 It is essential that an antifreeze mixture is retained in the cooling system at all times to act as a corrosion inhibitor and to protect the engine against freezing in winter months. The mixture should be made up from clean water with a low lime content (preferably rainwater) and a good quality ethylene glycol based antifreeze which contains a corrosion inhibitor and is suitable for use in aluminium engines.

2 The proportions of antifreeze to water required will depend on the maker's recommendations, but the mixture must be adequate to give protection down to approximately -30°C (-22°F).

3 Before filling with fresh antifreeze drain, and if necessary flush, the cooling system, as described in Sections 3 and 4. Check that all hoses are in good condition and that all clips are secure, then fill the system, as described in Section 5.

4 The antifreeze should be renewed every two years to maintain adequate corrosion protection. Do not use engine coolant antifreeze in the windscreen or tailgate wash systems; it will damage the car's paintwork and smear the glass. *Finally remember that antifreeze is poisonous and must be handled with due care.*

5 In climates which render frost protection redundant, it is still necessary to use a corrosion inhibitor in the cooling water. Suitable inhibitors should be available from a local GM agent or other reputable specialist.

10.6A Fitting a new O-ring to the water pump

10.6B Fitting the water pump to the block (engine removed)

12 Fault diagnosis – cooling system

Symptom	Reason(s)
Overheating	Coolant level low
	Radiator blocked (internally or externally)
	Kinked or collapsed hose
	Thermostat defective
	Fan or thermoswitch defective
	Engine out of tune
	Cylinder head gasket blown, or head or block cracked
	New engine not yet run-in
	Exhaust system partly blocked
	Engine oil low or incorrect grade
	Brakes binding
Overcooling	Faulty, incorrect or missing thermostat
Loss of coolant	Loose hose clips
	Hoses perished or leaking
	Radiator or heater matrix leaking
	Filler/pressure cap defective
	Cylinder head gasker blown, or head or block cracked

Chapter 3 Fuel and exhaust systems

For modifications, and information applicable to later models, see Supplement at end of manual

Contents

Air cleaner – element renewal	4	General description	1
Cold start cable – removal and refitting	14	Glow plugs – removal and refitting	12
Exhaust system – general	17	Idle speed – checking and adjustment	5
Fault diagnosis – fuel system	18	Idle stop solenoid – removal and refitting	9
Fuel filter – element renewal	3	Maintenance and inspection	2
Fuel injection pump – removal, overhaul and refitting	8	Manifolds – removal and refitting	16
Fuel injection pump timing – checking and adjustment	7	Maximum speed – checking and adjustment	6
Fuel injectors – removal, overhaul and refitting	10	Preheating system – description and testing	11
Fuel tank – removal and refitting	15	Throttle cable – removal and refitting	13

Specifications

General
System type	Rear-mounted fuel tank, combined lift and injection pump, self-priming system
Firing order	1 – 3 – 4 – 2 (No 1 at pulley end)
Fuel filter type	Champion L113
Air filter type	Champion U503

Fuel
Fuel type	Commercial Diesel fuel (DERV)
Fuel tank capacity	Between 42 and 61 litres (9.2 and 13.4 gallons) according to model

Injection pump
Maker's identification:
16D engine (early models)	VE 2300 R 82
16D engine (later models)	VE 4/9 F 2400 RTV 8253
16DA engine	VE 4/9 F 2300 R 215
Drive	From camshaft drivebelt
Identification of No 1 cylinder union	D

Injectors
Maker's identification:
16D engine (early models)	DN 05D 193
16D engine (later models)	DN 5D 193
16DA engine	Flat pintle type

Opening pressure:
16D engine (new)	140 to 148 bar (2030 to 2146 lbf/in^2)
16D engine (used)	135 bar (1958 lbf/in^2)
16DA engine	135 bar (1958 lbf/in^2)

Glow plugs
Make and type	Bosch 0250 200 056, Champion CH-68, or equivalent
Rating	11V, 90W

Adjustment data
Idle speed	825 to 875 rpm
Governed maximum speed	5600 rpm
Injection commencement at idle	3° to 5° BTDC

Pump timing setting (see text, Section 7):
16D engine	1 ± 0.05 mm (0.039 ± 0.002 in)
16DA engine	0.9 ± 0.05 mm (0.035 ± 0.002 in)

Torque wrench settings

	Nm	lbf ft
Injection pump and brackets:		
Main bracket to block	25	18
Subsidiary brackets – M6 bolts	14	10
Subsidiary brackets – M8 bolts	25	18
Injection pump sprockets bolts	25	18
Injectors to head	70	52
Glow plugs	40	30
Manifolds to head	22	16

Chapter 3 Fuel and exhaust systems

1 General description

The fuel supply and injection system follows normal practice for modern passenger and light commecial Diesel vehicles. A combined lift and injection pump, driven from the camshaft drivebelt, draws fuel from the tank and distributes it to each cylinder in turn. The injectors deliver a high pressure spray of fuel into the swirl chambers, where combustion starts. Excess fuel from the pump and the injectors is returned to the tank. A filter, which also acts as a water trap, protects the pump from contaminated fuel.

Cold starting is assisted by pre-heating the combustion chambers electrically. This system is automatically controlled – see Section 11. A driver-operated cold start control, similar to a manual choke control on petrol engines, causes the pump to deliver extra fuel and alters the injection timing slightly to improve cold start performance.

Unlike some older systems, manual bleeding or venting is not necessary even if the fuel tank is run dry. Provided that the battery is in good condition, simply cranking the engine on the starter motor will eventually bleed the system. (The starter motor should not be operated for more than ten seconds at a time. Allow five seconds between periods of operation.)

The fuel injection system is inherently robust and reliable. If the specified maintenance is carried out conscientiously it should give little trouble. Some components can only be overhauled or repaired by specialists; the home mechanic is warned against attempting operations beyond those described in this Chapter, unless qualified to do so.

Warning: *It is necessary to take certain precautions when working on the fuel system components, particularly the fuel injectors. Before carrying out any operations on the fuel system, refer to the precautions given in Safety First! at the beginning of this manual and to any additional warning notes at the start of the relevant Sections*

2 Maintenance and inspection

1 Cleanliness is important for trouble-free operation, and this includes clean fuel. Only use reputable fuel from proper dispensing pumps. Fuel from unknown cans or bowsers should be avoided unless steps can be taken to remove dirt and water from it. Do not use marine, agricultural or aviation fuel.

2 During cold weather it may be necessary to use fuel additives to prevent waxing of the fuel. Winter grade fuel in the UK is typically protected down to approx -9°C (+16°F). If temperatures are expected to fall below this point, a proprietary additive should be used with the fuel. Such additives are available at filling stations; **do not** use paraffin (kerosene) or petrol (gasoline), which may be illegal and/or dangerous. Obviously the additive must be present in the fuel lines before the temperature falls to the critical level. Wax formation usually affects the fuel filter first; in an emergency, pouring hot water over the outside of the filter may succeed in melting the wax. Do not pour hot water on the injection pump, however.

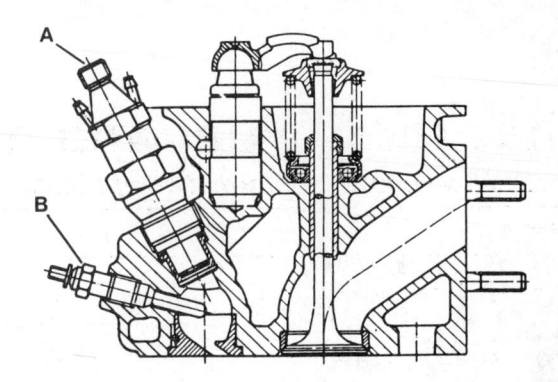

3 At every engine oil change the fuel filter should be drained to remove any water which may have accumulated. Place a container underneath the filter drain plug. Slacken the vent plug (on top of the filter carrier) by one turn (photo). Slacken the drain plug (on the filter base) by one turn and allow the water layer to drain from the filter (photo). When clean fuel emerges, tighten both screws and remove the container. Dispose of the water/fuel layer safely.

4 Renew the fuel filter element at the intervals specified in Routine Maintenance. See Section 3.

5 At the same intervals, or more frequently in dusty conditions, renew the air cleaner element. See Section 4.

6 Regularly inspect the fuel lines and fuel tank breather hoses for security and good condition. Rectify leaks without delay.

7 Also inspect the exhaust system regularly for security and freedom from corrosion.

Fig. 3.1 Sectional view of cylinder head through injector and glow plug (Sec 1)

A Injector   B Glow plug

2.3A Slackening the fuel filter vent plug

2.3B Fuel filter drain plug (arrowed)

Chapter 3 Fuel and exhaust systems

8 Renew the glow plugs at the intervals specified in Routine Maintenance. See Section 12.

9 Check the idle speed at the specified intervals, or whenever it seems to be incorrect. See Section 5.

10 If the engine suffers from excessive smoking, knocking or loss of performance which cannot be cured by procedures described in this Chapter, consult a GM dealer or other reputable specialist without delay.

3 Fuel filter – element removal

1 Thoroughly clean the filter element and carrier, especially around the joint between the two.

2 Drain the fuel from the filter by opening the vent and drain plugs. Catch the fuel in a container and dispose of it safely; tighten the vent plugs.

3 Unscrew the filter element with the aid of a chain or strap wrench, similar to that used for oil filter removal (photo). Discard the old element and make sure that no sealing rings have been left behind on the filter carrier.

4 Place the sealing ring supplied with the new filter over the centre hole in the filter and clip it into place with the retainer (Figs. 3.2 and 3.3).

5 Lubricate the outer sealing ring with a smear of clean fuel. Offer the filter to the carrier and tighten it with firm hand pressure, or as directed by the filter manufacturer. Tighten the drain plug.

6 Start the engine and run it at a fast idle for a minute or so to vent the system. Check around the filter seal for leaks; tighten a little further if necessary.

4 Air cleaner – element renewal

1 Remove the air cleaner snorkel, which is held to the bonnet lock platform by two screws (photos).

2 Slacken the screws which tension the two retaining clips (photo).

3 Unhook the clips and remove the air cleaner box complete with element (photo).

4 Remove and discard the old element. Wipe clean inside the air cleaner box.

5 Fit the new element to the box, engaging the rubber seal round the edge (photo).

6 Engage the clips and tension them by tightening the screws.

7 Refit and secure the snorkel.

3.3 Fuel filter element removal

Fig. 3.2 Fuel filter sealing ring and retainer (both arrowed) (Sec 3)

Fig. 3.3 Fuel filter sealing ring and retainer (arrowed) in position (Sec 3)

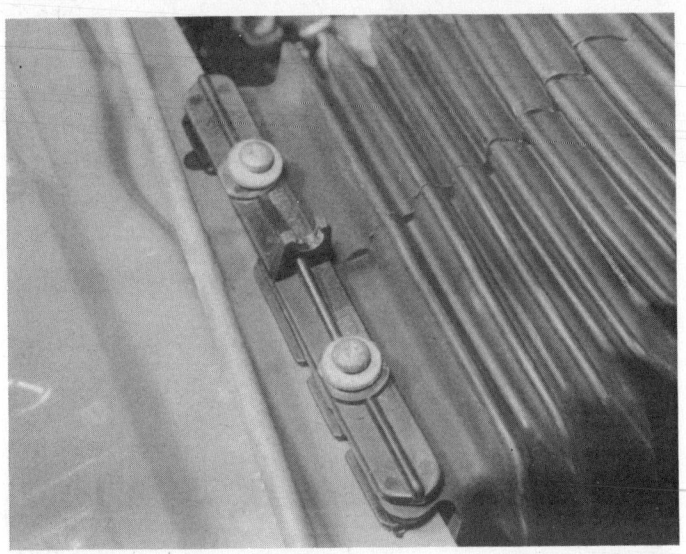

Chapter 3 Fuel and exhaust systems

73

4.1A Air cleaner snorkel securing screws

4.1B Removing the air cleaner snorkel

4.2 Air cleaner clip tensioning screw

4.3 Air cleaner clips released

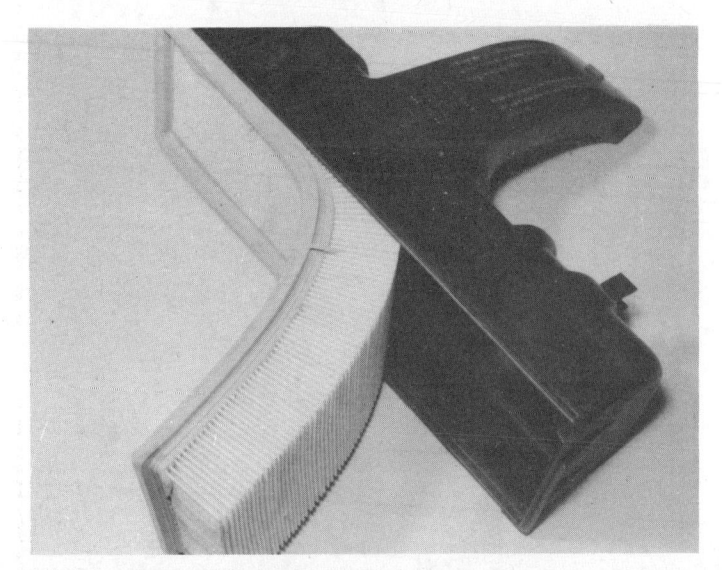

4.5 Fitting the new air cleaner element

5 Idle speed – checking and adjustment

Caution: *Keep clear of the cooling fan when making adjustments*

1 The main difficulty in making engine speed adjustments is the measurement of the speed. Conventional tachometers cannot be used because they are triggered by ignition system pulses. Proprietary instruments are available which operate by sensing the passage of (for instance) a chalk mark on the crankshaft pulley, but they are expensive.

2 If the relationship of road speed to engine rpm is known for any gear, the vehicle can be positioned with its front wheels off the ground and the speedometer reading converted to rpm. Apply the handbrake and chock the rear wheels securely if this method is adopted. The accuracy of the speedometer may not be great, especially at the low speeds involved at idle.

3 A third possibility is the use of a dynamic timing light (stroboscope) connected to the ignition system of a petrol-engined vehicle. Make a chalk or paint mark on the crankshaft pulley of the Diesel engine and shine the stroboscope at it. Run the petrol engine at the

Chapter 3 Fuel and exhaust systems

desired speed. When the Diesel engine is running at the same speed, the pulley chalk mark will appear stationary. (The same applies at half or twice the speed, so some common sense must be used).

4 If adjustment is necessary, remove the air cleaner snorkel to improve access. Slacken the locknut on the idle speed adjusting screw and turn the screw clockwise to increase the speed, anti-clockwise to decrease it. Tighten the locknut, without altering the screw position, when adjustment is correct (photos). Refit the snorkel.

6 Maximum speed – checking and adjustment

Caution: *Keep clear of the cooling fan when making adjustments*

1 There should not normally be any need to adjust the maximum speed, except after major component renewal.

2 Refer to Section 5 for ways of measuring engine speed. It is probably unwise to use the speedometer method, since there is a grave risk of injury or damage should anything go wrong.

3 Start the engine and gradually increase its speed, observing the tachometer or its equivalent, until the governed maximum speed is reached. (Do not accelerate the engine much beyond the specified maximum, should maladjustment make this possible).

4 If adjustment is necessary, remove the air cleaner snorkel to improve access. Slacken the locknut and turn the maximum speed adjuster screw until the desired result is obtained, then tighten the locknut without moving the screw (photo). Refit the snorkel.

5 Seal the screw with paint or thread locking compound if it is wished to guard against unauthorised adjustment of the maximum speed.

7 Fuel injection pump timing – checking and adjustment

1 Pump timing should only be necessary when fitting a new or overhauled pump, or if the timing is suspected of being wrong, or if the timing belt has been re-tensioned or renewed. A dial test indicator, with a long probe and a suitable support, will be needed.

2 The procedure as shown here was carried out during engine rebuilding. With the engine in the vehicle, it will be necessary to remove the drivebelt covers, the air cleaner snorkel and the clutch/flywheel cover.

3 Check the valve timing as decribed in Chapter 1, Section 6.

4 Bring the engine to TDC, No 1 firing. The timing mark on the pump sprocket must be aligned with the pip on the pump bracket (photo).

5 Turn the engine against the normal direction of rotation so that the flywheel TDC mark is approximately 5 cm (2 in) away from the TDC pointer.

6 Remove the central plug from the rear of the injection pump (photo).

7 Mount the dial test indicator with its probe entering the central plug hole. Zero the indicator.

8 Be prepared for fuel spillage during subsequent operations. (The makers specify the use of a probe which screws into, and presumably seals, the plug hole).

9 Bring the engine back to TDC, No 1 firing. When the timing marks are aligned, the dial test indicator should show a lift corresponding to the desired timing setting – see Specificatons (photo).

5.4A Idle speed adjusting screw (arrowed) on fuel injection pump

5.4B Adjusting the idle speed

6.4 Adjusting the maximum speed

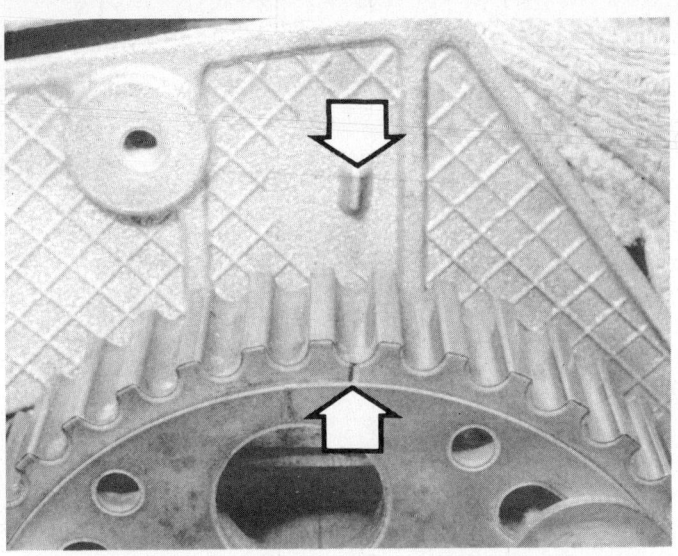

Chapter 3 Fuel and exhaust systems

7.4 Pump sprocket timing mark and pipe in alignment (arrows)

7.6 Removing the plug from the rear of the pump

7.9 Dial test indicator mounted with its probe in the plug hole

10 If adjustment is necessary, slacken the three bolts which clamp together the two halves of the pump sprocket (photo). Turn the inner part of the sprocket anti-clockwise (against the normal direction of rotation) as far as the slots will allow. The fit between the two parts of the sprocket is tight; a rod or soft metal drift may be needed to encourage the inner part to move.

11 With the sprocket positioned as just described and the engine still at TDC, No 1 firing, the dial test indicator should again read zero. Reset it if necessary.

12 Turn the inner part of the sprocket clockwise until the dial test indicator shows the desired lift, then tighten the sprocket clamp bolts.

13 Repeat the checking procedure from paragraph 5.

14 When the injection timing is correct, remove the test gear and refit the plug to the rear of the pump.

15 Refit the drivebelt covers and other disturbed components.

7.10 Slackening a sprocket clamping bolt

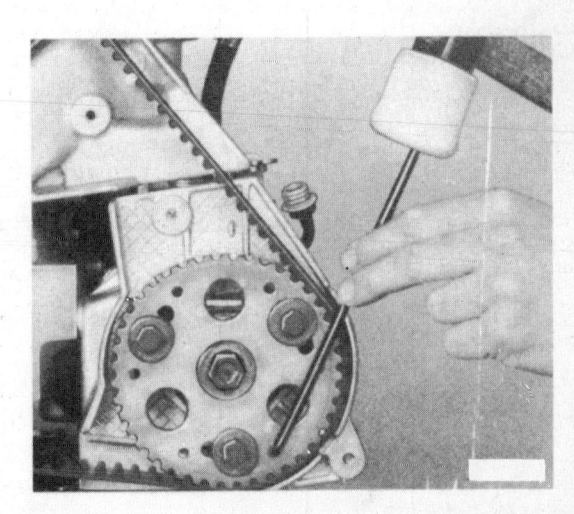

Fig. 3.4 Using a rod and mallet to move the inner part of the fuel pump sprocket (Sec 7)

8 Fuel injection pump – removal, overhaul and refitting

1 Slacken the camshaft drivebelt as described in Chapter 1, Section 5, and slip it off the pump sprocket.

2 Remove the air cleaner snorkel.

3 Disconnect the fuel feed and return hoses from the pump. Also disconnect the injected fuel return hose from the T-piece (photos). Be prepared for fuel spillage; plug or cap the open unions (eg with polythene and rubber bands) to keep fuel in and dirt out.

4 Clean around the unions, then remove the injection pipes. Do not separate the pipes from their brackets. Again, plug the open unions.

5 Disconnect the lead from the idle stop solenoid.

6 Disconnect the throttle and cold start cables from the pump as described in Sections 13 and 14.

7 Restrain the pump sprocket from turning and remove the central securing nut (photo).

8 Remove the sprocket from the pump shaft. The use of a puller is recommended (photo). If a 'face' puller is used, it will have to be secured with the bolts which clamp the two parts of the sprocket together; if these bolts have been disturbed, the pump timing will have to be reset after refitting. Make alignment marks between the parts of the sprocket if wished.

9 Recover the Woodruff key from the shaft if it is loose (photo).

10 Remove the two bolts which secure the sprocket end of the pump to the bracket. Access to the bolt on the engine side is achieved with a socket and a long extension (photo).

11 Remove the two bolts which secure the spring/damper bracket to the fuel pump bracket. Remove the pump complete with subsidiary brackets.

8.3A Fuel pump feed hose (A) and return hose (B) connections

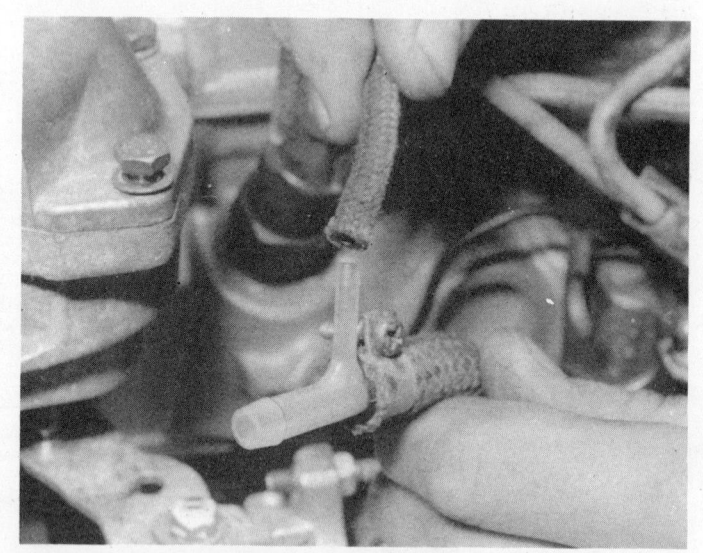

8.3B Disconnecting the injector fuel return hose

8.7 Undoing the pump sprocket nut

8.8 Using a puller to remove the pump sprocket

Chapter 3 Fuel and exhaust systems

8.9 Pump shaft and Woodruff key

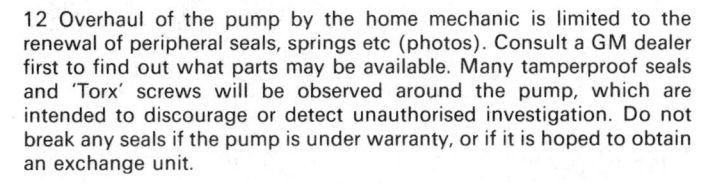

8.10 Removing a fuel pump securing bolt

12 Overhaul of the pump by the home mechanic is limited to the renewal of peripheral seals, springs etc (photos). Consult a GM dealer first to find out what parts may be available. Many tamperproof seals and 'Torx' screws will be observed around the pump, which are intended to discourage or detect unauthorised investigation. Do not break any seals if the pump is under warranty, or if it is hoped to obtain an exchange unit.

13 If a new pump is being fitted, transfer the brackets and other necessary components to it. Do not tighten the bracket bolts yet.

14 Refit in the reverse order of removal, noting the following points:
   (a) Slacken the subsidiary bracket bolts, then tighten them in the sequence shown in Fig. 3.5
   (b) Tighten all fastenings to the specified torque, when known
   (c) Check the valve timing as described in Chapter 1, Section 6
   (d) Check the injection timing as described in Section 7 if the sprocket was separated, or if a new pump has been fitted

8.12A Fuel pump shaft oil seal (arrowed)

8.12B Top view of fuel pump showing throttle lever and spring

8.12C Rear view of fuel pump. Union D supplies No 1 cylinder; arrow shows operating sequence

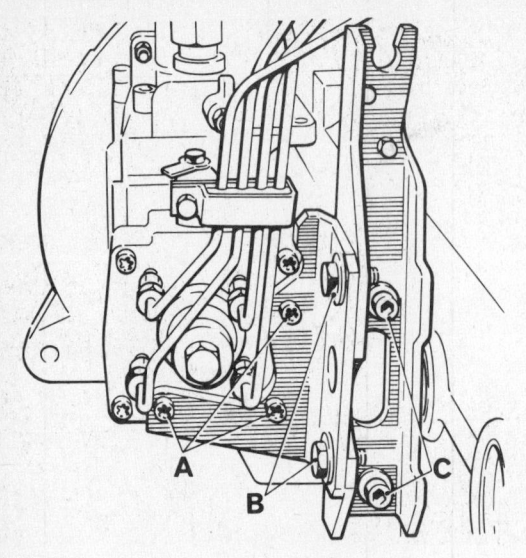

Fig. 3.5 Fuel pump bracket bolts – tighten in sequence A - B - C (Sec 8)

15 A good deal of cranking on the starter motor will be required to prime the pump before the engine will run. Do not operate the starter for more than 10 seconds at a time, then pause for a 5 second period to allow the battery and starter motor to recover.

16 Check all fuel unions for leaks when the engine is running, and again when it has been stopped.

9 Idle stop solenoid – removal and refitting

1 The idle stop solenoid is energised all the time that the engine is running; when the 'ignition' is switched off, the solenoid cuts off the fuel supply at the pump and the engine stops.

2 To remove the solenoid, first disconnect the battery earth lead.

3 Disconnect the electrical lead from the screw terminal on top of the solenoid (photo).

4 Clean around the solenoid, then unscrew it from the pump.

5 Recover the O-ring spring and plunger (photo). Cover the pump orifice to keep dirt out.

6 Commence refitting by inserting the plunger and spring into the orifice (photo).

7 Fit and tighten the solenoid, using a new O-ring. Do not overtighten. Reconnect the solenoid and the battery.

10 Fuel injectors – removal, overhaul and refitting

Warning: *Exercise extreme caution when working on the fuel injectors. Never expose the hands or any part of the body to injector spray, as working pressure can cause the fuel to penetrate the skin with possibly fatal results. You are strongly advised to have any work which involves testing the injectors under pressure carried out by a dealer or fuel injection specialist*

1 Remove the air cleaner snorkel.

2 Clean around the unions, then remove the injection pipes from the injectors and fuel pump. Plug or cap the open unions on the pump. Also remove the fuel return hoses from the injectors. Be prepared for fuel spillage.

3 Clean around the bases of the injectors, then unscrew and remove them (photo). A deep socket spanner, size 27 mm (1 1/16 in AF), is the

9.3 Idle stop solenoid terminal (arrowed)

9.5 Idle stop solenoid, O-ring, spring and plunger

9.6 Fitting the plunger and spring

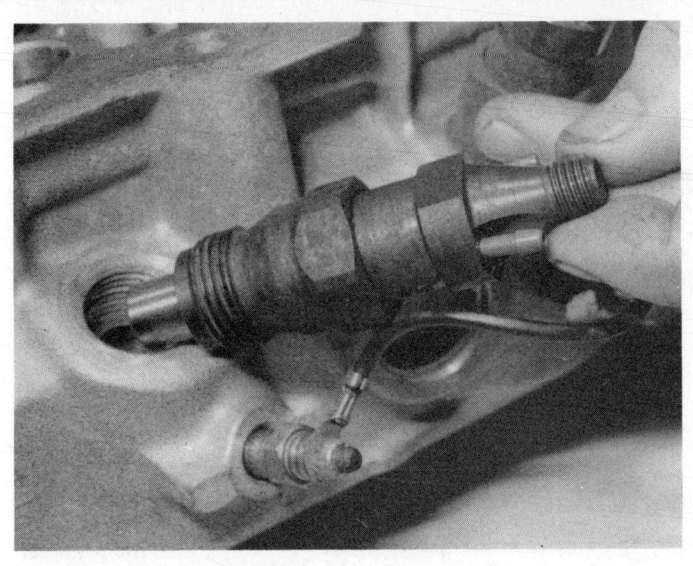

Chapter 3 Fuel and exhaust systems 79

10.3 Removing a fuel injector

10.4A Removing the large plain washer ...

10.4B ... and the small corrugated washer

Fig. 3.6 Correct installation of injector small washer – arrow points to cylinder head (Sec 10)

best tool to use. Open-ended, ring or even adjustable spanners may do at a pinch if there is room to use them. On later models (January 1987 on), due to the alteration of the location of the crankcase vent hose connection, access to No 2 cylinder fuel injector is restricted. If a socket wrench is used, its diameter may require reducing by grinding.

4 Recover the injector washers. There are two per injector: the large one seals the injector carrier-to-head joint, the small one seals the injector tip (photos). Obtain new washers for reassembly.

5 Injectors can be overhauled, but this is not a DIY job. Consult a GM dealer or other reputable specialist.

6 Commence refitting by inserting the small washers. Note that they must be fitted the right way up (Fig. 3.6).

7 Fit the large washers, either way up, then screw in the injectors. Tighten the injectors to the specified torque.

8 Refit the return hoses and the injection pipes.

9 Run the engine and check that there are no leaks at the pipe unions. Check again with the engine stopped.

11 Preheating system – description and testing

1 The swirl chambers are heated immediately before start-up by electrical heater plugs (usually called glow plugs). When the key is turned to the 'on' position, a warning light illuminates to inform the driver that preheating is in progress. When the light goes out, the engine is ready to be started.

2 Switching of the high current needed to heat the glow plugs is carried out by a relay mounted on the bulkhead (photo) or on one of the suspension turrets. The relay also determines the time for which the current needs to be applied.

3 If malfunction of the system is suspected, check first that battery voltage appears at the glow plug bus bar for a few seconds when the key is first turned to the 'on' position. If not, there is a fault in the wiring or the relay. Testing of the relay is by substitution.

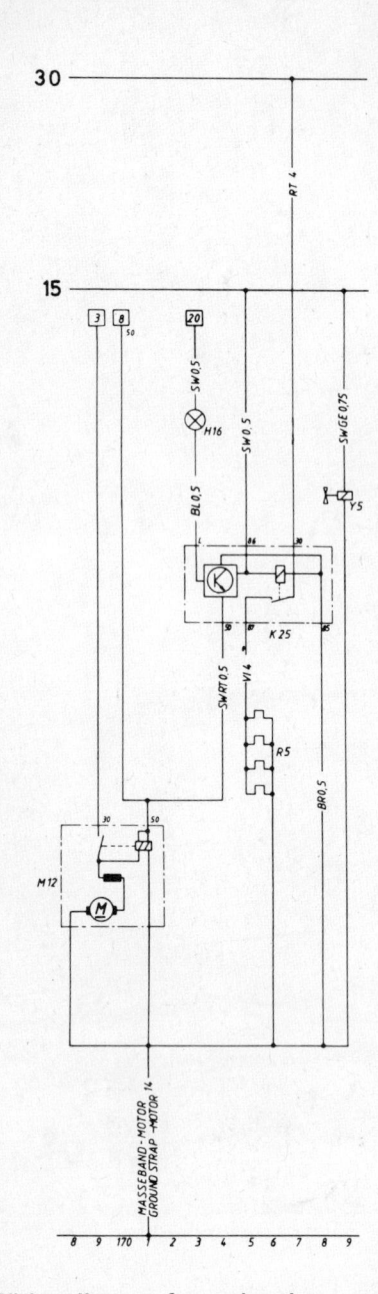

Chapter 3 Fuel and exhaust systems

11.2 Glow plug relay unplugged

surge, so if the reading is much above or below 32 to 36 amps, there is a defect in one or more plugs. Disconnect each plug in turn to isolate the culprit.

5 In the absence of an ammeter, a 12 volt test lamp can be used. Remove the bus bar and connect the lamp between the battery live (+) terminal and each glow plug in turn. If the lamp lights, the glow plug is probably OK (or there is a short-circuit). If the lamp does not light, the glow plug is defective.

12 Glow plugs – removal and refitting

1 Disconnect the battery earth lead.

2 Remove the air cleaner snorkel.

3 Disconnect the feed wire from the bus bar (photo).

4 Unscrew the nuts and remove the bus bar and link wire from the glow plugs (photo). Note the disposition of the washers.

5 Unscrew and remove the glow plugs (photo).

6 Refit in the reverse order of removal. Tighten the glow plugs to the specified torque and make sure that the electrical connections are clean and tight.

13 Throttle cable – removal and refitting

1 Prise the cable inner ball end fitting off its stud on the pump throttle lever (photo). Access is tight: it will be improved by removing the throttle damper, which is secured by similar ball and socket fittings.

2 Free the cable outer grommet from the bracket on the pump (photo).

3 Inside the vehicle, compress the spring and release the cable from the 'keyhole' fitting on the pedal arm (photo).

4 Free the cable from any clips or ties and remove it from under the bonnet.

Fig. 3.7 Wiring diagram for preheating system (typical) (Sec 11)

H16	Warning light	8	From starter switch
K25	Glow plug relay	15	Live rail (ignition controlled)
M12	Starter motor		
R5	Glow plugs	20	From instrument fuse
Y5	Idle stop solenoid	30	Live rail (full-time)
3	From battery		

Colour code
BL Blue   RT Red
BR Brown   SW Black
GE Yellow   VI Violet

4 If battery voltage is present at the bus bar but one or more glow plugs do not seem to be working, it is possible to identify the defective plug(s) with the aid of a high range ammeter (say 0 to 50A). (An ohmmeter is unlikely to be able to distinguish between the resistance of a good plug – less than 1 ohm – and a short circuit). Connect the ammeter between the bus bar and the feed wire. Have an assistant turn the key. Each plug will draw between 8 and 9 amps after an initial

12.3 Glow plug bus bar feed wire

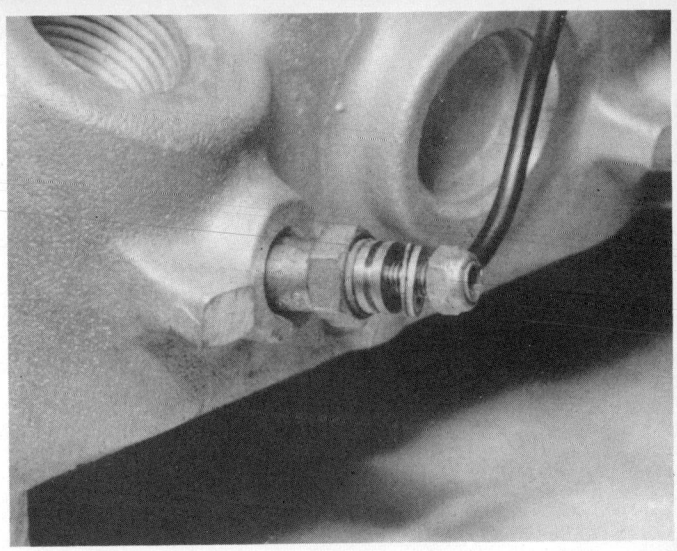

12.4 No 1 glow plug showing link wire connection

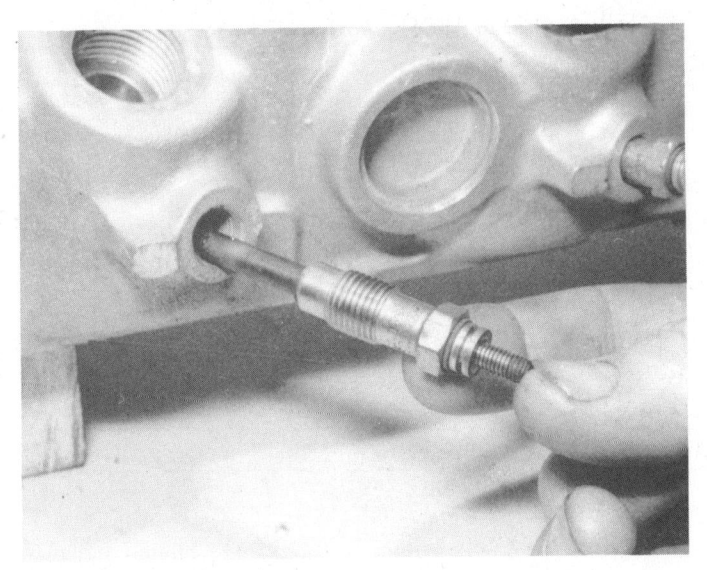

12.5 Removing a glow plug

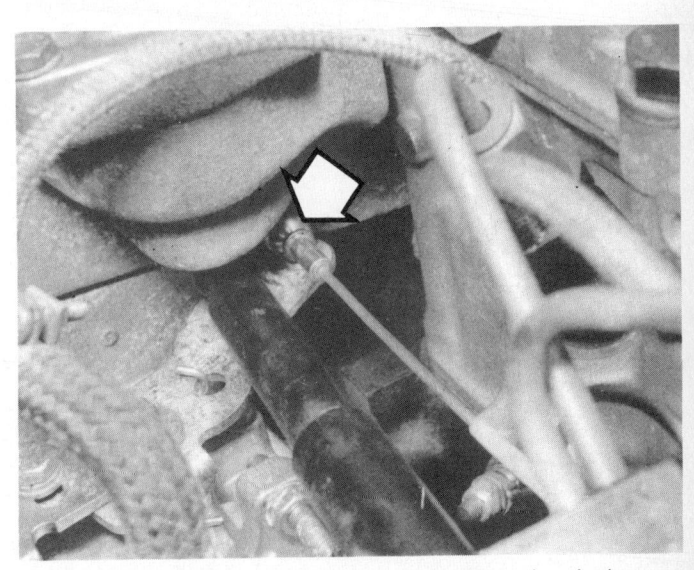

13.1 Throttle cable ball end fitting (arrowed) next to throttle damper

13.2 Throttle cable outer grommet (arrowed) next to other end of throttle damper

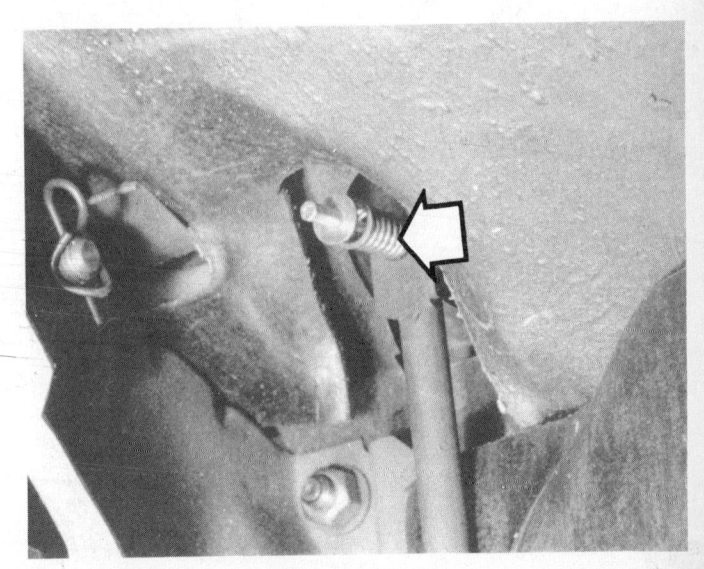

13.3 Throttle cable fitting at pedal end

Chapter 3 Fuel and exhaust systems

5 Refit in the reverse order of removal. Adjust the positon of the clip on the pump end of the cable outer so that with the pedal released, there is a small amount of slack in the cable inner.

6 Remember to refit the throttle damper if it was removed.

14 Cold start cable – removal and refitting

1 Free the cable inner at the pump end by releasing the spring clip which links it to the operating rod. Access is poor, but it is possible to release the clip by feel (photos).

2 Remove the clip which secures the cable outer grommet in its bracket.

3 Inside the vehicle, remove the cold start control knob by pressing out its roll pin. Undo the nut which secures the cable outer to the facia panel.

4 Withdraw the cable into the vehicle, removing additional interior trim if necessary. Refer to the appropriate manual for petrol-engined models.

5 Refit in the reverse order of removal. There is no provision for cable adjustment.

15 Fuel tank – removal and refitting

1 The procedure is essentially the same as for petrol-engined vehicles, with the bonus that it is much less dangerous. The following points are also to be noted.

2 Inspect the vent hoses closely (photo) and renew them if they have become porous or are in any way suspect. It is possible for water to enter the tank through deteriorated vent hoses. It is important that the vent hose which carries the water deflector is correctly fitted. The hose should be inserted into the water deflector as shown in Fig. 3.8. Failure to do this could cause collapse of the fuel tank under certain conditions.

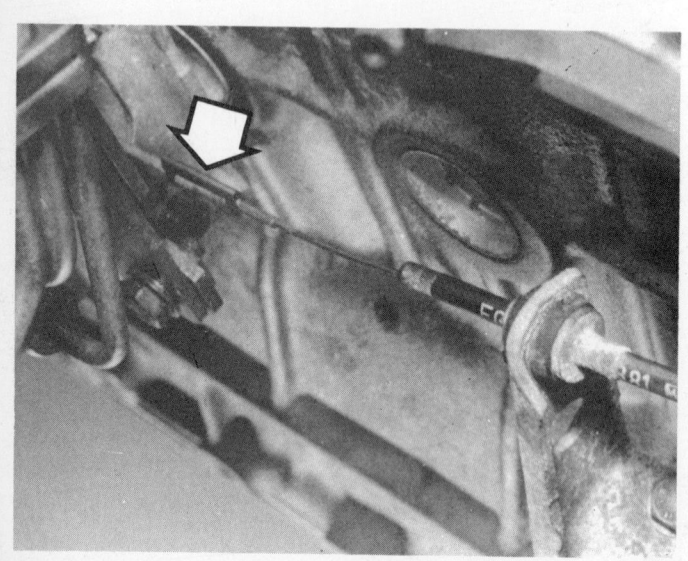

14.1A Cold start cable – spring clip arrowed

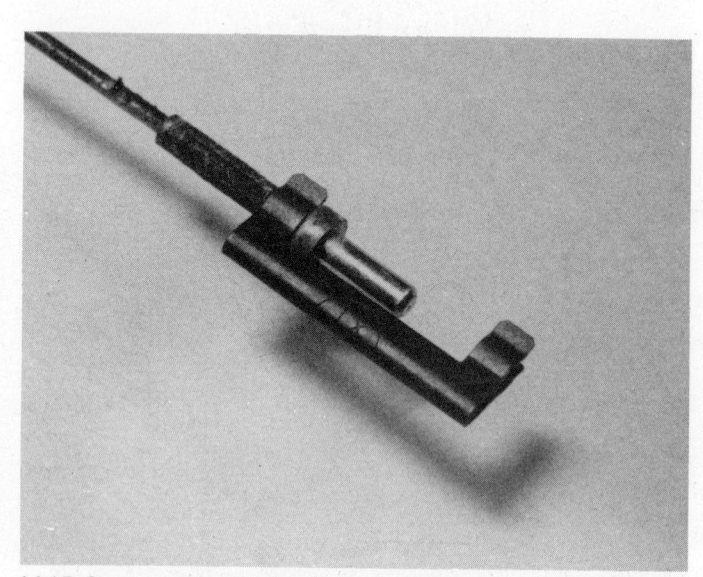

14.1B Close-up showing detail of spring clip

15.2 Fuel tank vent hoses

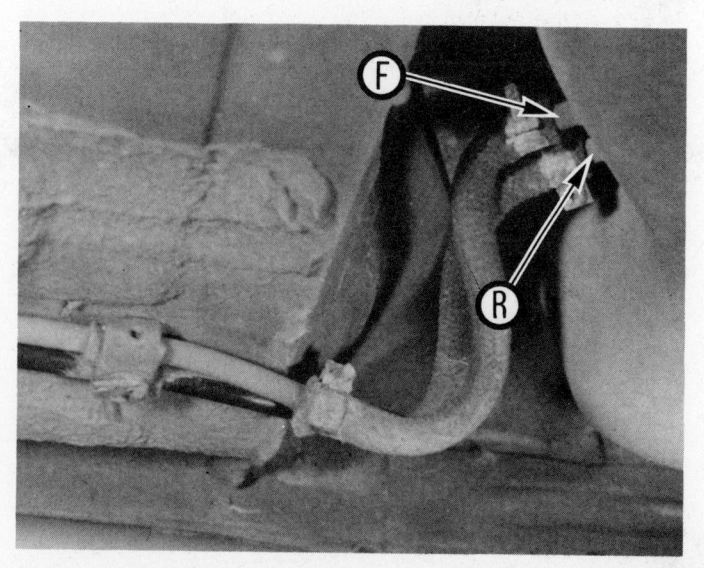

15.3 Fuel tank flow (F) and return (R) connections

Chapter 3 Fuel and exhaust systems

3 Do not confuse the fuel flow (suction) and return connections. From 1984 model year the flow connection on the tank is marked with a blue sleeve and the flow pipe is coloured blue. In case of doubt, note that the flow connection is nearest the centre-line of the vehicle, while the return connection is nearer the right-hand side (photo).

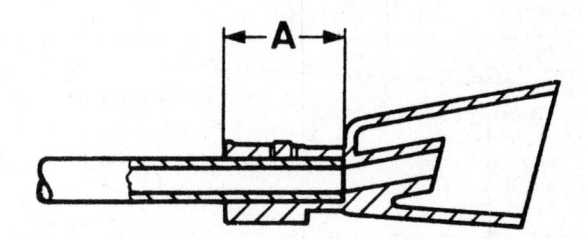

Fig. 3.8 Vent hose correctly fitted into water deflector (Sec 15)

A = at least 22.0 mm (0.87 in)

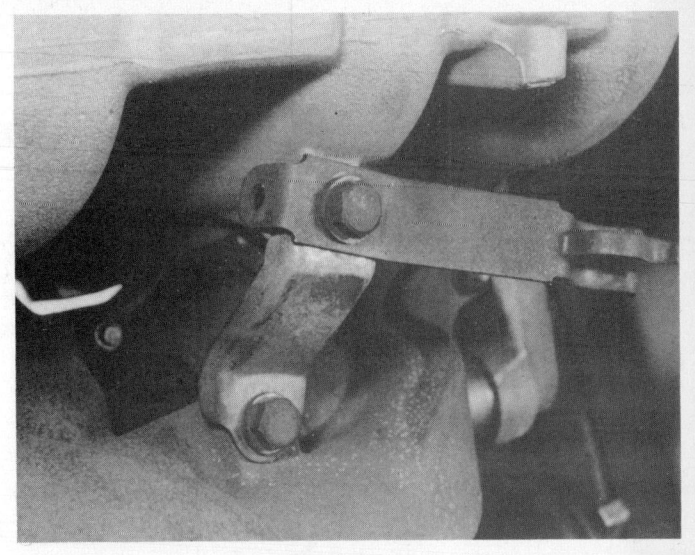

16.4 Manifold link and clutch cable bracket

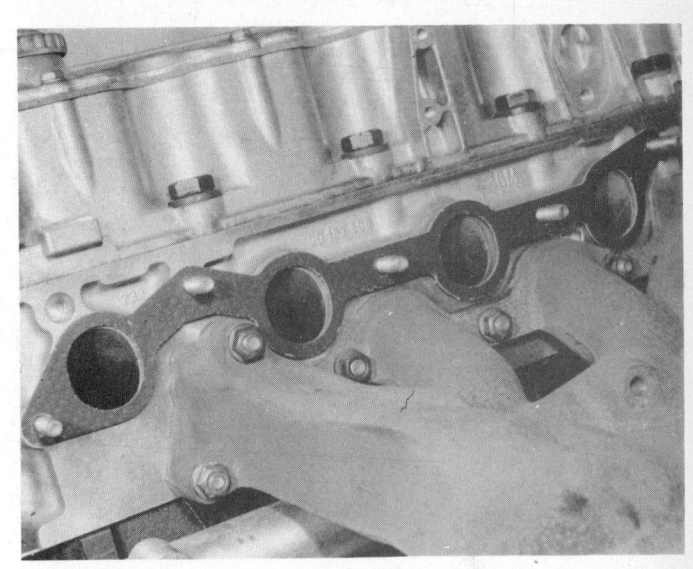

16 Manifolds – removal and refitting

Inlet manifold

1 Disconnect the battery earth lead.

2 Remove the air cleaner as described in Section 4.

3 Disconnect the breather hose from the camshaft cover. (Note that the other end of this hose is attached to a distribution tube inside the manifold).

4 Remove the bolts which secure the inlet-to-exhaust manifold link. Recover the link and the clutch cable bracket (photo).

5 Remove the five nuts which secure the manifold to the head. Withdraw the manifold and recover the gasket (photo).

6 Refit in the reverse order of removal. Use a new gasket and tighten the nuts evenly to the specified torque. Do not forget the clutch cable bracket.

Exhaust manifold

7 Remove the inlet manifold as just described.

8 Remove the exhaust downpipe and recover the flange gasket.

9 Remove the seven nuts which secure the manifold to the head. Withdraw the manifold and recover the gasket (photo).

10 Use new gaskets on both joints when refitting, and apply some anti-seize compound to the stud threads.

11 Tighten the manifold-to-head nuts evenly to the specified torque, then refit the exhaust downpipe.

12 Refit the inlet manifold.

16.5 Inlet manifold removal leaving gasket

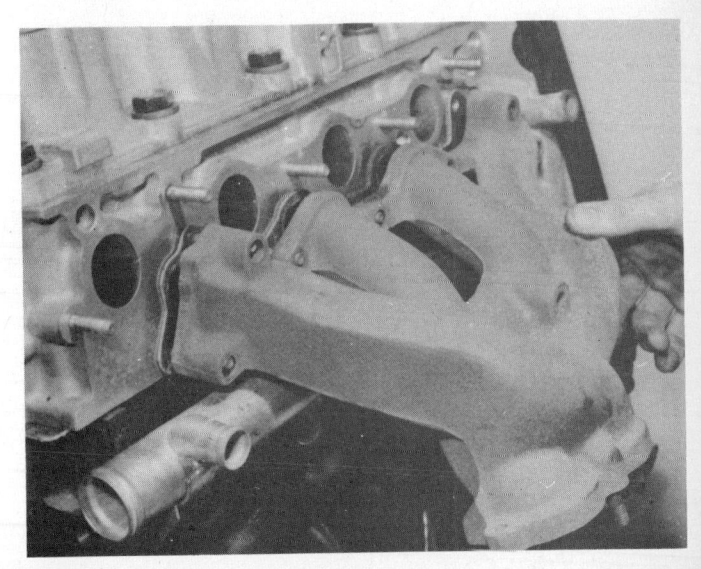

17 Exhaust system – general

The exhaust system is similar in construction and mounting to that fitted to petrol-engined models. Refer to the appropriate manual for petrol-engined vehicles.

16.9 Removing the exhaust manifold

18 Fault diagnosis – fuel system

Faults in the fuel injection system can cause knocking and other engine noises which to the inexperienced ear suggest imminent mechanical failure. When such noises are caused by one particular injector locking or sticking, the faulty injector can be located by slackening the injector pipe unions at each injector in turn with the engine idling. The noise will disappear when the union on the defective injector is slackened.

Symptom	Reason(s)
Uneven running or misfiring	Air leak in fuel feed line Water in fuel Injector seating washers leaking Injectors defective Injection timing incorrect Injection pump defective
Engine stalls	Idle speed adjustment incorrect Injection timing incorrect Filter sealing ring incorrectly fitted Injection pump defective
Lack of power	Fuel filter clogged Fuel tank vent blocked Exhaust system restricted Air cleaner blocked Throttle cable wrongly adjusted Fuel feed or return pipes restricted or connections interchanged at tank Injection timing incorrect Injector(s) defective Maximum speed adjustment incorrect
Abnormal knocking	Injector(s) defective Injector washers missing Injector insulating sleeve burnt Injector wrongly assembled Injection timing incorrect Injection pump defective
Engine loses power and stops	Fuel tank empty Air leak in fuel feed line Fuel filter clogged Fuel tank vent blocked Excessive use of cold start device
Black smoke in exhaust	Air cleaner clogged Injector seating washers leaking Injectors defective Excessive use of cold start device Injection timing incorrect Injection pump defective

Chapter 4 Clutch and transmission

Contents

General description 1
Maintenance and overhaul 2

Specifications

Clutch

Type	Single dry plate, diaphragm spring, cable-operated
Driven plate diameter	200, 203 or 216 mm (7.9, 8.0 or 8.5 in), according to model
Adjustment:	
Pedal free play	Nil
Pedal stroke	138 to 145 mm (5.4 to 5.7 in)

Manual transmission

Type	Four or five forward speeds and one reverse; integral final drive	
Maker's designation:		
Astra/Kadett up to mid-1986, and all Ascona/Cavalier	F16/4 or F16/5	
Astra/Kadett from mid-1986	F13/4 or F13/5	
	F16	**F13**
Ratios (typical):		
1st	3.42 : 1	3.55 : 1
2nd	1.95 : 1	1.96 : 1
3rd	1.28 : 1	1.30 : 1
4th	0.89 : 1	0.89 : 1
5th (when applicable)	0.71 : 1	0.71 : 1
Reverse	3.33 : 1	3.31 : 1
Final drive ratio (F16 and F13)	3.74 : 1 or 3.94 : 1	
Lubricant type/specification	Gear oil, viscosity SAE 80 to GM 4753 M (Duckhams Hypoid 80W/90)	

Automatic transmission

Type	Three forward speeds and one reverse; integral final drive
Maker's designation	125 THM
Ratios (typical):	
1st	2.84 : 1
2nd	1.60 : 1
3rd	1.00 : 1
Reverse	2.07 : 1
Primary drive	1.12 : 1
Final drive	3.74 : 1
Torque converter multiplication	2.4 : 1 max
Lubricant type/specification	Dexron II type ATF to GM 6137 M (Duckhams Uni-Matic)

1 General description

Only manual transmission is available in passenger vehicles equipped with the GM Diesel engine in the UK. In other territories automatic transmission is optionally available, and in the UK automatic transmission can be specified for the light vans.

Manual transmission has four or five forward speeds, all with synchromesh, and one reverse. It is fitted in conjunction with a single dry plate clutch. The final drive/differential unit is integral with the rest of the transmission and shares the same lubricant.

Automatic transmission has three forward speeds and one reverse. Changing between the forward speeds is normally fully automatic, the actual change points being influenced by roadspeed and throttle position. The driver can lock out top and intermediate gears when required for particular conditions.

Drive to the front wheels is by open tubular driveshafts of unequal length, with constant velocity joints at each end.

2 Maintenance and overhaul

Maintenance and overhaul procedures are as described in the appropriate manual for petrol-engined vehicles. Overhaul procedures for the F13 series transmission will be incorporated in the appropriate manual when available.

Chapter 5 Braking system

Contents

Description and maintenance 1
Vacuum pump – removal and refitting 3
Vacuum pump – testing 2

Specifications

System type .... As for petrol-engined models, but with vacuum for servo assistance developed by pump

Vacuum pump
Drive .... Direct from camshaft
Lubrication .... From engine lubrication system
Test values:
    Vacuum at 2750 rpm .... 0.75 bar (22.2 in Hg)
    Residual vacuum 30 seconds after stopping .... 0.73 bar (21.6 in Hg) minimum

Brake fluid type/specification .... Hydraulic fluid to GME 05301 or GM 4653 M (Duckhams Universal Brake and Clutch Fluid)

Torque wrench setting   Nm   lbf ft
Vacuum pump to camshaft housing 28   21

1 Description and maintenance

The braking system is in most respects identical to that fitted to the equivalent petrol-engined models. Because of the Diesel engine's different manifold vacuum characteristics, however, it is necessary to provide an alternative source of vacuum for the brake servo. This source takes the form of a vacuum pump, driven directly from the camshaft and taking the place occupied by the distributor on spark ignition engines.

Routine maintenance of the vacuum pump is not required; braking system maintenance is therefore as described for petrol-engined models.

2 Vacuum pump – testing

1 If there is a lack of servo assistance and the vacuum pump is suspect, it can be tested as follows.

2 Disconnect the vacuum pipe from the pump and connect a vacuum gauge in its place.

3 Run the engine and check that the required vacuum is developed at the appropriate speed (see Specifications).

4 Stop the engine and check that the vacuum is maintained at the specified level. Obviously the gauge and its fittings must be leak-tight.

5 Disconnect the vacuum gauge and refit the servo vacuum pipe.

6 A defective vacuum pump must be renewed. Do not attempt to dismantle the pump: some of the end cover screws are sealed to detect unauthorised entry.

3 Vacuum pump – removal and refitting

1 Disconnect the servo vacuum pipe from the pump. Do this by counterholding the large union nut and unscrewing the small one (photo).

2 Remove the two Allen screws and withdraw the pump from the camshaft housing (photos). Be prepared for some oil spillage.

3 Recover the oil pipe and the driving dog.

4 Use new sealing rings on the oil pipe and on the pump spigot when refitting.

5 Refit the oil pipe and the driving dog to the pump (photos).

6 Offer the pump to the camshaft housing, making sure that the dog teeth engage with the slot in the camshaft. Fit the Allen screws and tighten them to the specified torque.

7 Reconnect and secure the vacuum pipe.

3.1 Disconnect the vacuum pipe from the pump

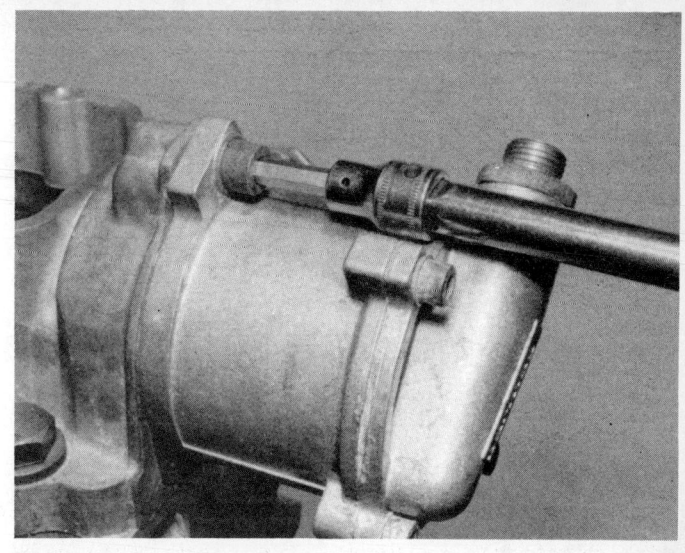

3.2A Undoing a vacuum pump securing screw

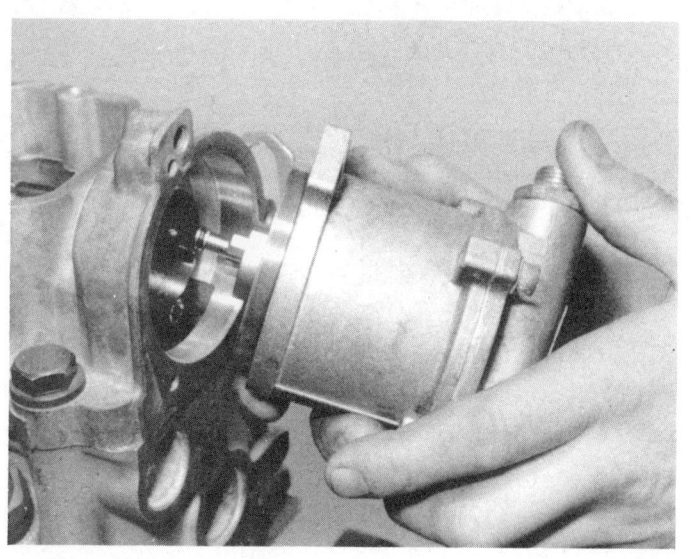

3.2B Removing the vacuum pump

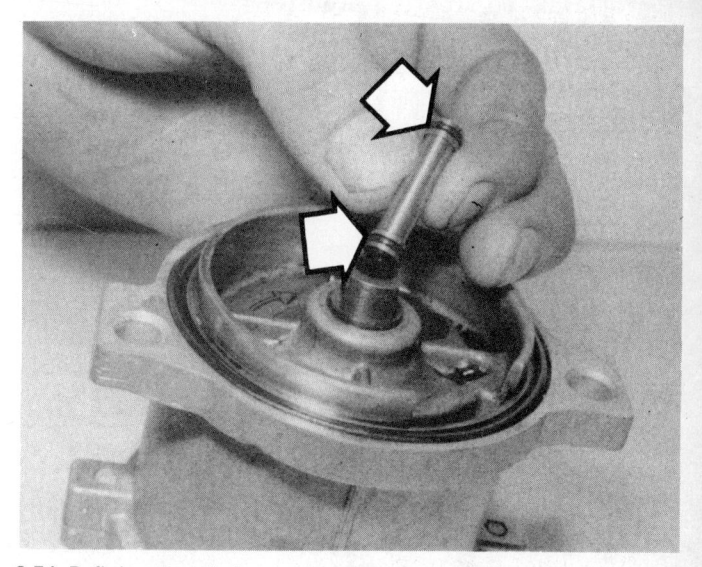

3.5A Refitting the oil pipe – note sealing rings (arrowed)

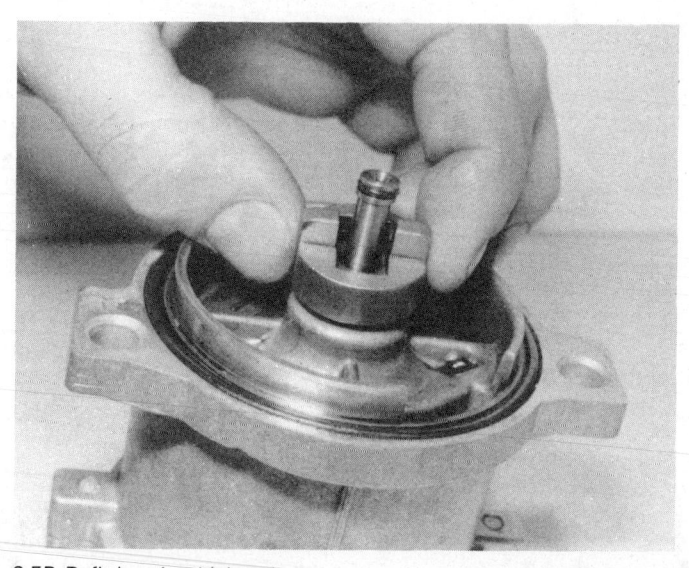

3.5B Refitting the driving dog

Chapter 6 Suspension and steering

For information applicable to later models, see Supplement at end of manual

Contents

General description 1	Wheels and tyres – general care and maintenance 2

Specifications

Power-assisted steering fluid type .... Dexron II type ATF (Duckhams Uni-Matic)

Tyre pressures (cold) in bar (lbf/in^2)*

	Light load		Medium load		Full load	
	Front	Rear	Front	Rear	Front	Rear
Astra/Kadett Saloon/Hatchback (up to 1984):						
155 SR 13	1.9 (28)	1.9 (28)	–	–	2.1 (31)	2.4 (35)
175/70 SR 13	1.8 (26)	1.8 (26)	–	–	2.0 (29)	2.3 (33)
Astra/Kadett Estate (up to 1984):						
155 SR 13	1.9 (28)	2.2 (32)	2.0 (29)	2.5 (36)	2.2 (32)	3.2 (46)
175/70 SR 13	1.8 (26)	2.0 (29)	1.9 (28)	2.4 (35)	2.0 (29)	2.8 (41)
175/65 SR 14	1.9 (28)	1.9 (28)	2.0 (29)	2.4 (35)	2.1 (31)	2.9 (42)
Astra/Kadett Saloon/Hatchback (1985 to 1991):						
155 SR 13	1.8 (26)	1.6 (23)	–	–	1.9 (28)	2.4 (35)
All other tyre sizes	1.8 (26)	1.6 (23)	–	–	1.9 (28)	2.1 (31)
Astra/Kadett Estate/Van (1985 to 1991, except Astramax)	1.9 (28)	1.9 (28)	2.0 (29)	2.4 (35)	2.0 (29)	2.8 (41)
Astramax Van:						
155 SR 13	2.0 (29)	2.2 (32)	–	–	2.1 (31)	3.0 (44)
165 SR 13	1.8 (26)	2.2 (32)	–	–	1.8 (26)	3.0 (44)
Belmont	2.0 (29)	1.8 (26)	–	–	2.1 (31)	2.5 (36)
Cavalier/Ascona (except Estate):						
Pre-1987	1.9 (28)	1.7 (25)	–	–	2.0 (29)	2.2 (32)
1987 to 1988	1.9 (28)	1.7 (25)	–	–	2.1 (31)	2.3 (33)
Cavalier/Ascona Estate	2.0 (29)	2.0 (28)	–	–	2.4 (35)	2.4 (35)

* Tyre pressures are taken from the maker's technical data current at the time of writing; pressures specified in vehicle handbook may differ. If in doubt consult a GM dealer or tyre specialist

1 General description

On most vehicles covered by this book, suspension is independent at the front and semi-independent at the rear. On the larger Vans, rear suspension is by beam axle and leaf springs. Steering is by rack-and-pinion, with power assistance available as an optional extra.

Tyre pressures tend to be slightly different on Diesel models because of the increased nose weight – see Specifications. Otherwise, maintenance and overhaul procedures are as described in the appropriate manual for petrol-engined vehicles.

2 Wheels and tyres – general care and maintenance

Wheels and tyres should give no real problems in use provided that a close eye is kept on them with regard to excessive wear or damage. To this end, the following points should be noted.

Ensure that tyre pressures are checked regularly and maintained correctly. Checking should be carried out with the tyres cold and not immediately after the vehicle has been in use. If the pressures are checked with the tyres hot, an apparently high reading will be obtained owing to heat expansion. Under no circumstances should an attempt

Chapter 6 Suspension and steering

be made to reduce the pressures to the quoted cold reading in this instance, or effective underinflation will result.

Underinflation will cause overheating of the tyre owing to excessive flexing of the casing, and the tread will not sit correctly on the road surface. This will cause a consequent loss of adhesion and excessive wear, not to mention the danger of sudden tyre failure due to heat build-up.

Overinflation will cause rapid wear of the centre part of the tyre tread coupled with reduced adhesion, harsher ride, and the danger of shock damage occurring in the tyre casing.

Regularly check the tyres for damage in the form of cuts or bulges, especially in the sidewalls. Remove any nails or stones embedded in the tread before they penetrate the tyre to cause deflation. If removal of a nail *does* reveal that the tyre has been punctured, refit the nail so that its point of penetration is marked. Then immediately change the wheel and have the tyre repaired by a tyre dealer. Do *not* drive on a tyre in such a condition. In many cases a puncture can be simply repaired by the use of an inner tube of the correct size and type. If in any doubt as to the possible consequences of any damage found, consult your local tyre dealer for advice.

Periodically remove the wheels and clean any dirt or mud from the inside and outside surfaces. Examine the wheel rims for signs of rusting, corrosion or other damage. Light alloy wheels are easily damaged by 'kerbing' whilst parking, and similarly steel wheels may become dented or buckled. Renewal of the wheel is very often the only course of remedial action possible.

The balance of each wheel and tyre assembly should be maintained to avoid excessive wear, not only to the tyres but also to the steering and suspension components. Wheel imbalance is normally signified by vibration through the vehicle's bodyshell, although in many cases it is particularly noticeable through the steering wheel. Conversely, it should be noted that wear or damage in suspension or steering components may cause excessive tyre wear. Out-of-round or out-of-true tyres, damaged wheels and wheel bearing wear/maladjustment also fall into this category. Balancing will not usually cure vibration caused by such wear.

Wheel balancing may be carried out with the wheel either on or off the vehicle. If balanced on the vehicle, ensure that the wheel-to-hub relationship is marked in some way prior to subsequent wheel removal so that it may be refitted in its original position.

General tyre wear is influenced to a large degree by driving style – harsh braking and acceleration or fast cornering will all produce more rapid tyre wear. Interchanging of tyres may result in more even wear, but this should only be carried out where there is no mix of tyre types on the vehicle. However, it is worth bearing in mind that if this is completely effective, the added expense of replacing a complete set of tyres simultaneously is incurred, which may prove financially restrictive for many owners.

Front tyres may wear unevenly as a result of wheel misalignment. The front wheels should always be correctly aligned according to the settings specified by the vehicle manufacturer.

Legal restrictions apply to the mixing of tyre types on a vehicle. Basically this means that a vehicle must not have tyres of differing construction on the same axle. Although it is not recommended to mix tyre types between front axle and rear axle, the only legally permissible combination is crossply at the front and radial at the rear. When mixing radial ply tyres, textile braced radials must always go on the front axle, with steel braced radials at the rear. An obvious disadvantage of such mixing is the necessity to carry two spare tyres to avoid contravening the law in the event of a puncture.

In the UK, the Motor Vehicles Construction and Use Regulations apply to many aspects of tyre fitting and usage. It is suggested that a copy of these regulations is obtained from your local police if in doubt as to the current legal requirements with regard to tyre condition, minimum tread depth, etc.

Chapter 7 Electrical system

Contents

Alternator – removal and refitting 4	Starter motor – removal and refitting 6
Alternator brushes – renewal 5	Starter motor (direct drive type) – dismantling and reassembly 8
Alternator drivebelt – inspection, renewal and adjustment 3	Starter motor (reduction gear type) – dismantling and reassembly 7
Battery – maintenance 2	
General description 1	Wiring diagrams – general 9

Specifications

General
System type .... 12V, negative earth. As for petrol-engined models except where noted below

Battery
Capacity (original equipment) .... 66 Ah
Cold start current .... 400 A
Test load current .... 270 A
Voltage after 15 seconds at test load:
   At -18°C (0°F) 8.5
   At -10°C (14°F) 8.8
   At 0°C (32°F) 9.1
   At 10°C (50°F) 9.4
   At or above 21°C (70°F) 9.6

Starter motor
Make and type .... Bosch GF (direct drive) or DW (reduction gear)
Rating .... 12V, 1.7 kW
Commutator minimum diameter:
   Type GF .... 33.5 mm (1.32 in)
   Type DW .... 31.2 mm (1.24 in)
Brush length, minimum:
   Type GF .... 13.0 mm (0.5 in)
   Type DW .... 8.0 mm (0.3 in)

Alternator
Make .... Bosch or Delco
Maximum output .... 45, 55 or 65 A according to model and equipment
Brush wear limit:
   Bosch .... 5.0 mm (0.2 in) protrusion
   Delco .... 11.0 mm (0.4 in) overall length

Torque wrench settings
	Nm	lbf ft
Starter motor (pinion end) to block/bellhousing	45	33
Alternator bracket to block	40	30

1 General description

The electrical system is very similar to that found on petrol-engined models. The Diesel starter motor is of heavier construction, to deal with the greater demands of compression ignition; the battery and alternator are similarly uprated.

The preheating system (glow plugs and relay), which is electrically operated, is covered in Chapter 3.

Only a few basic procedures are given in this Chapter. For a fuller treatment refer to the appropriate manual for petrol-engined vehicles.

2 Battery – maintenance

1 Keep the battery terminals clean and free from corrosion. Any corrosion can be removed with an old toothbrush and a solution of sodium bicarbonate – do not let any solution get inside the cells.

2 Coat the battery terminals with a smear of petroleum jelly, or proprietary anti-corrosion product, when they are clean. Make sure that the terminal clamps are secure, but do not overtighten.

3 When removable cell covers are fitted, check at the specified intervals that electrolyte is covering the plates in every cell. If there are

Chapter 7 Electrical system

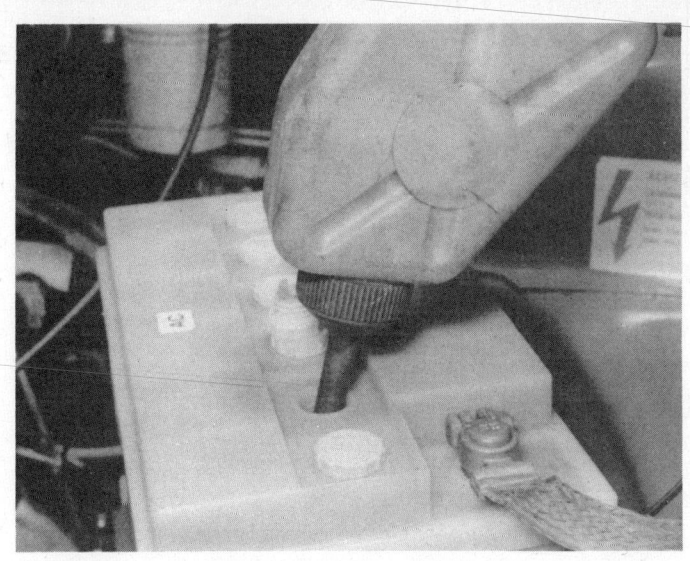

2.3 Topping up the battery with distilled water

Fig. 7.1 Charge condition indicator found on original equipment maintenance-free batteries. Green dot denotes full charge (Sec 2)

Darkened Indicator WITH GREEN DOT — Darkened Indicator NO GREEN DOT — **MAY BE JUMP STARTED**

Light or Bright Yellow Indicator NO GREEN DOT — **DO NOT JUMP START**

no level marks on the battery, make sure that the plates are covered to a depth of approximately 6 mm ($1/4$ in). Top up if necessary using distilled or de-ionized water (photo). Do not overfill, and mop up any spillage immediately.

4 If a battery hydrometer is available, it is instructive occasionally to check the state of charge of the battery. Use the hydrometer in accordance with its maker's instructions. Variation between cells is more significant than the overall state of charge. One cell reading which differs greatly from the others suggests either an incorrect electrolyte mix in that cell, or possible cell failure.

5 If it appears that acid has been lost from one or more cells at some time and the deficiency made good with water, consult a GM dealer or battery specialist. Do not attempt to mix or replenish electrolyte at home. Even in its dilute state, electrolyte is poisonous and corrosive.

6 Many batteries nowadays are of the 'maintenance-free' type, requiring no inspection or topping-up throughout their life. Sometimes they incorporate a charge indicator (Fig. 7.1). Follow the instructions on the battery case when dealing with such a battery: note particularly that rapid or 'boost' charging is usually forbidden due to the risk of explosion.

7 Charging of the battery from an external source should not normally be necessary. In extreme conditions (mainly short journeys with much electrical equipment in use), or to revive a flagging battery, assistance from a domestic charger may be useful. Disconnect the battery from the vehicle, and preferably remove it completely, before charging. The charge rate (in Amps) should not exceed one-tenth of the battery capacity (in Amp-hours), except under carefully controlled conditions. Remember that the battery gives off hydrogen gas, which is inflammable and potentially explosive during charging.

8 Inspect the battery tray occasionally for signs of corrosion. Neutralise and repaint corroded areas without delay.

3 Alternator drivebelt – inspection, renewal and adjustment

1 Periodically inspect the drivebelt for cracks, fraying, glazing or other signs of deterioration. Inspect the belt pulleys for shiny areas in the bottoms of the grooves – if fitting a new belt does not cure this, new pulleys are required. When properly fitted and adjusted, the belt transmits power through its side surfaces.

2 It is worth renewing the drivebelt as a precautionary measure at every fourth major service, even if it appears to be in good condition.

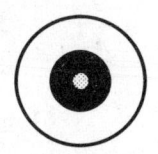

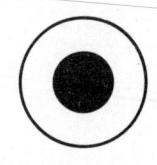

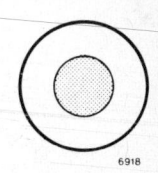

3 To remove the drivebelt, first slacken the alternator pivot and adjusting strap nuts and bolts.

4 Remove the steering pump drivebelt, when fitted.

5 Move the alternator towards the engine, slip the belt off the pulleys and remove it.

6 Fit the new belt over the pulleys and adjust it as follows.

7 Tighten the alternator fastening slightly, so that the alternator can just be moved by hand.

8 Move the alternator away from the engine until the belt tension is correct. In the absence of a special belt tension gauge, aim for a tension such that the belt can be deflected approximately 12 mm ($1/2$ in) by firm finger pressure in the middle of its run. If using a lever to move the alternator, only use a wooden or plastic one and only lever at the pulley end.

9 Tighten the alternator fastenings fully when the belt tension is correct.

10 Refit and tension the steering pump drivebelt, when applicable.

11 The tension of a new drivebelt should be rechecked after a few hundred miles.

4 Alternator – removal and refitting

1 Disconnect the battery earth (negative) lead (photo).

2 Remove the alternator drivebelt as described in Section 3. The belt can stay on the crankshaft pulley if wished.

3 Disconnect the leads from the rear of the alternator.

4 Unbolt the adjusting strap and earth link from the alternator.

5 Unbolt the alternator mounting bracket from the block and remove the alternator, complete with bracket, from below. When the bracket forms part of an engine mounting, that too will have to be disconnected – see Chapter 1, Section 14.

6 As an alternative to removing the alternator with its bracket, it can be removed alone from above if the manifolds are first removed. See Chapter 3, Section 16.

7 Refit in the reverse order of removal. Tighten the alternator bracket bolts to the specified torque and adjust the drivebelt as described in the previous Section.

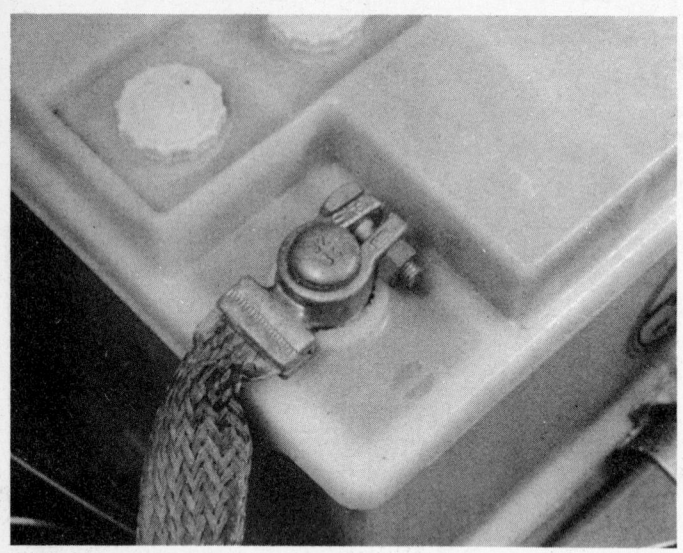

Chapter 7 Electrical system

4.1 Battery earth lead

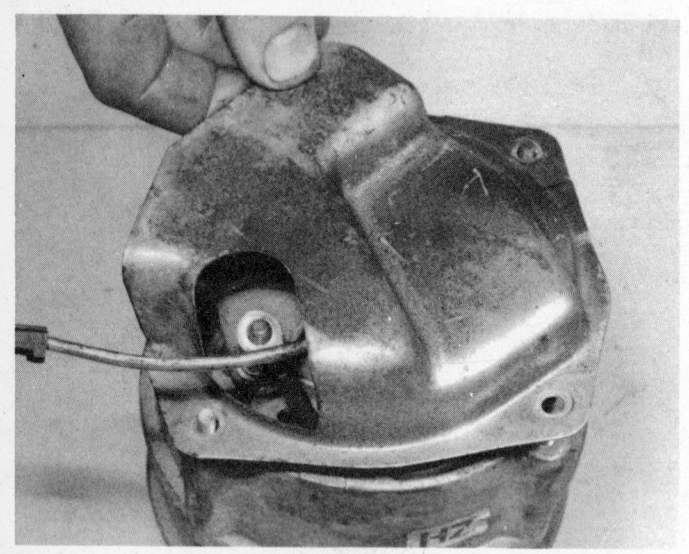

5.2 Removing the alternator heat shield

5.3 Undoing a brush carrier/voltage regulator screw

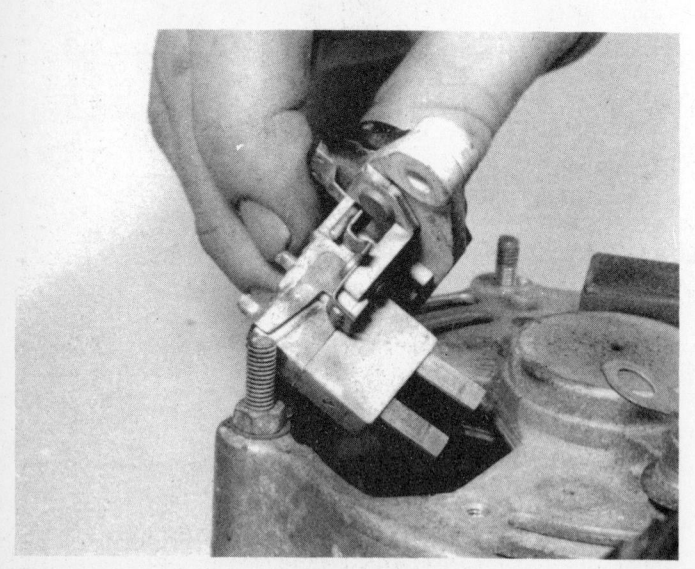

5.4 Removing the voltage regulator/brush carrier

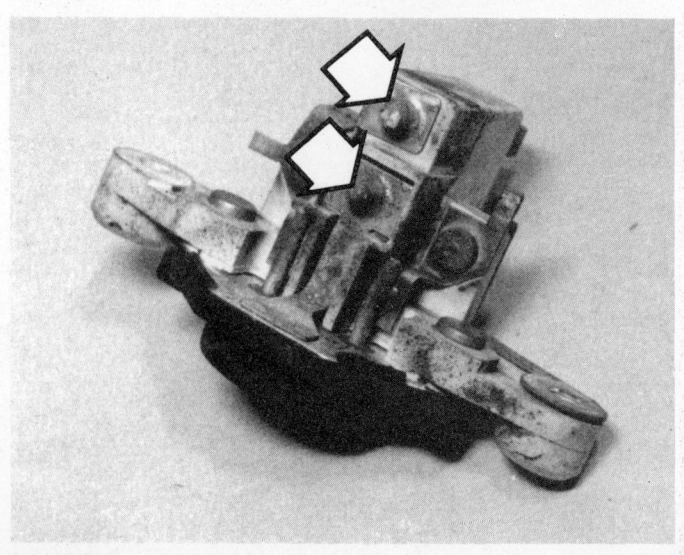

5.5 Brushes must be unsoldered at the points arrowed

5 Alternator brushes – renewal

1   Remove the alternator as described in the previous Section. (In theory the Bosch alternator can be dealt with in situ, but the access is not very good).

2   Remove the heat shield from the rear of the alternator (photo). It is secured by three nuts.

Bosch

3   Remove the two screws which secure the brush carrier/voltage regulator to the rear of the alternator (photo).

4   Remove the brush carrier/regulator assembly (photo).

5   Brush renewal entails the unsoldering of the old brush leads and the soldering in of the new ones at the points shown (photo). A certain familiarity with soldering technique is desirable if this is to be successful.

6   Make sure that the new brushes can move freely in their holders, then refit and secure the brush carrier/regulator.

Fig. 7.2 Exploded view of Bosch alternator (Sec 5)

1 Pulley nut
2 Pulley halves
3 Fan
4 Through-bolt
5 Drive end housing
6 Drive end bearing
7 Bearing retainer
8 Voltage regulator/brush carrier
9 Slip ring end housing
10 Rectifier
11 Stator
12 Slip ring end bearing
13 Rotor

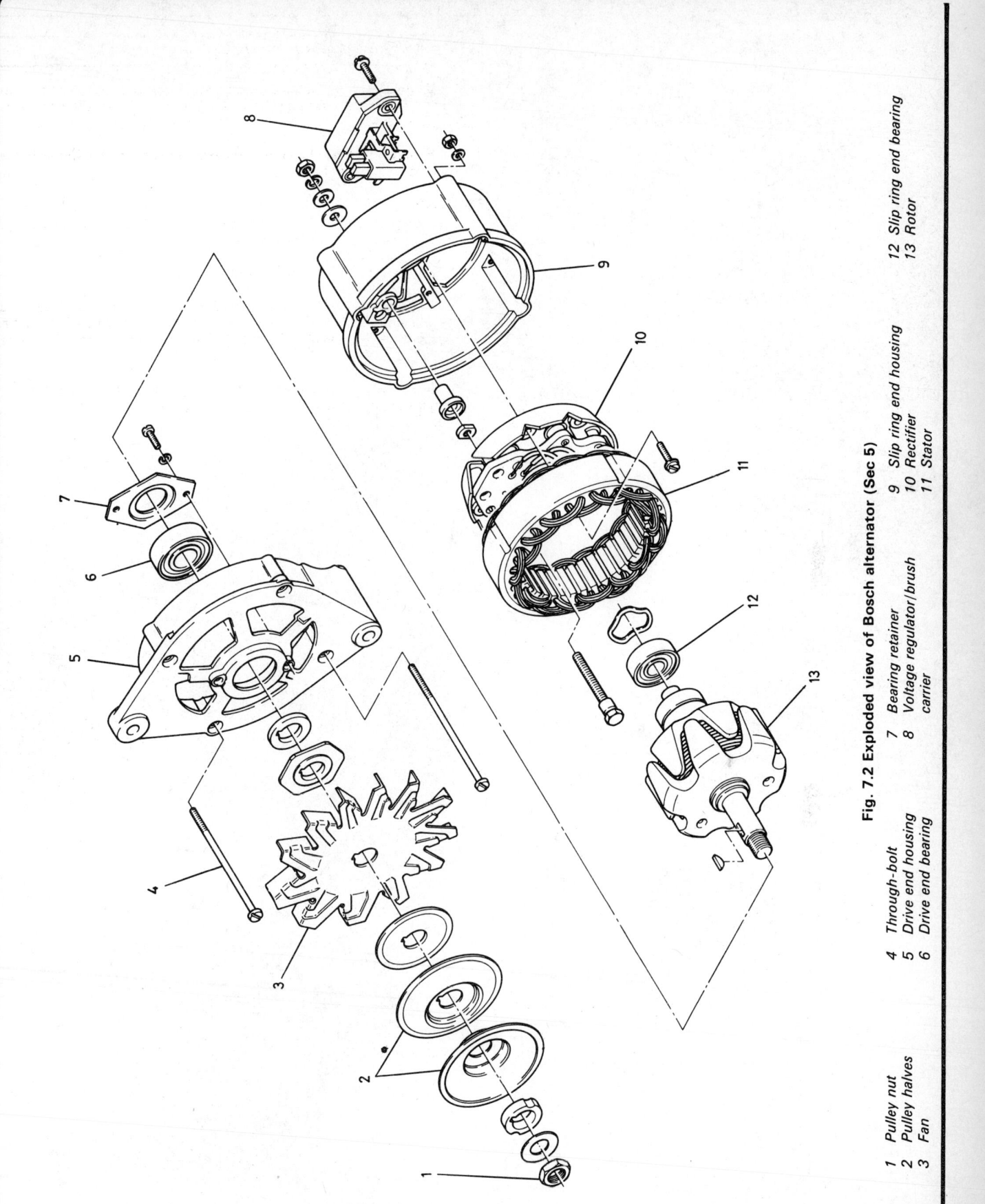

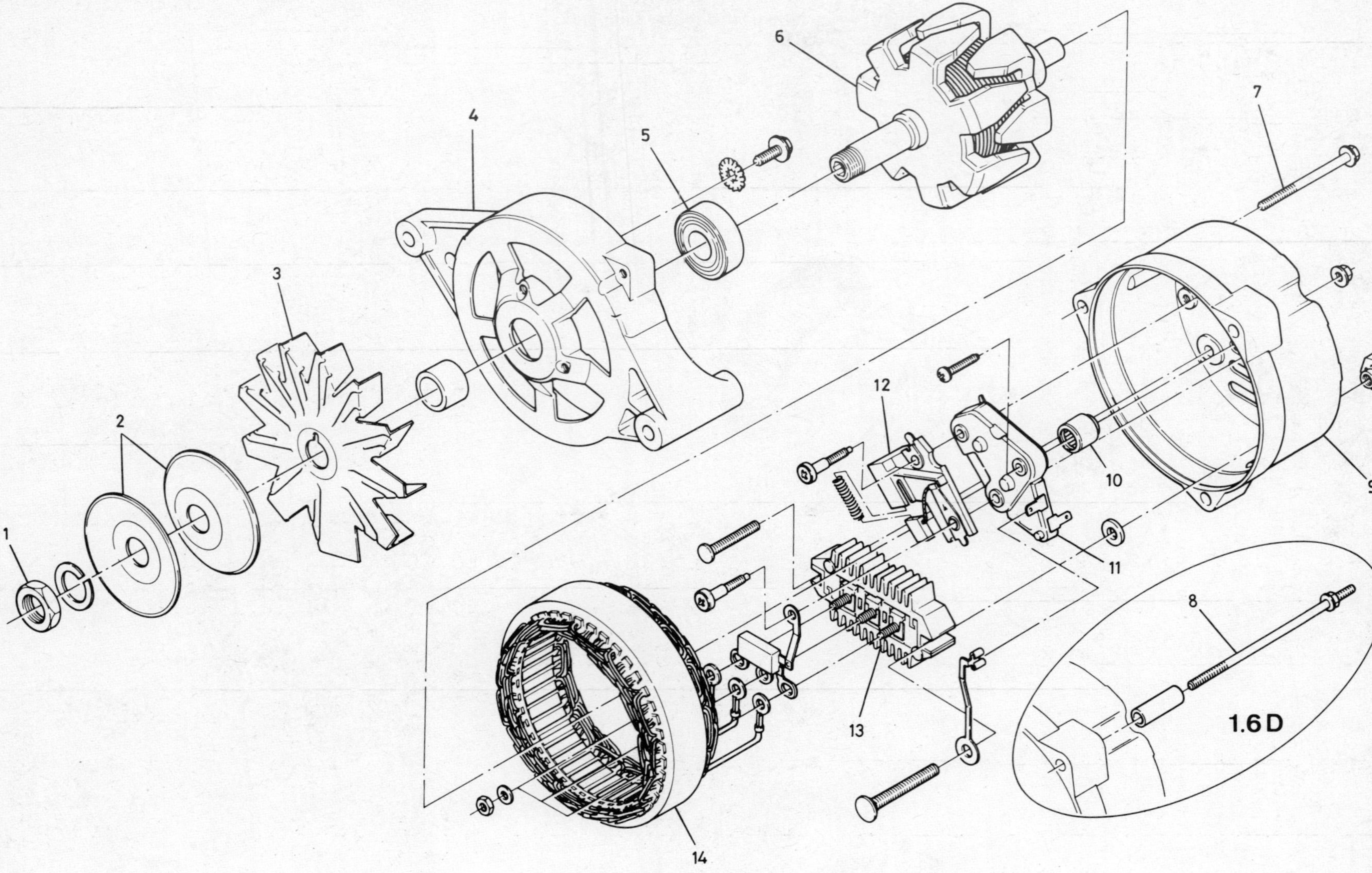

Fig. 7.3 Exploded view of Delco-Remy alternator (Sec 5)

1 Pulley nut
2 Pulley halves
3 Fan
4 Drive end housing
5 Drive end bearing
6 Rotor
7 Through-bolt (except Diesel)
8 Through-bolt and spacer (Diesel)
9 Slip ring end housing
10 Slip ring end bearing
11 Voltage regulator
12 Brush carrier
13 Rectifier
14 Stator

Chapter 7 Electrical system

Delco-Remy

7 Make alignment marks between the drive end housing and the slip ring end housing.

8 Undo the three through-bolts and separate the two housings (photo). Recover the spacers. Refer to Section 7, paragraph 4, if difficulty is experienced in undoing the through-bolts.

9 Check the condition of the rotor slip rings (photo). Clean them if necessary with a petrol-moistened cloth.

10 Remove the three nuts and washers which secure the stator leads to the rectifier (photo). Remove the stator, then undo the terminal screw and remove the rectifier.

11 Remove the two screws which secure the brush carrier and voltage regulator. Note the insulating washers under the screw heads.

12 Carefully unsolder the old brush leads and solder in the new ones in exactly the same place.

13 Make sure that the new brushes can move freely in their holders.

14 Refit the brush carrier and regulator. Push the brushes into their holders and retain them in the retracted position with a thin drill or a piece of stiff wire. One end of the wire or drill must protrude through the hole in the slip ring end housing (photo).

15 Refit the rectifier and secure the stator leads.

16 Assemble the two housings, observing the alignment marks previously made, and secure with the spacers and through-bolts.

17 Pull the wire or drill out of the hole so that the brushes drop into position (photo).

All types

18 Refit and secure the heat shield.

19 Refit the alternator as described in Section 4.

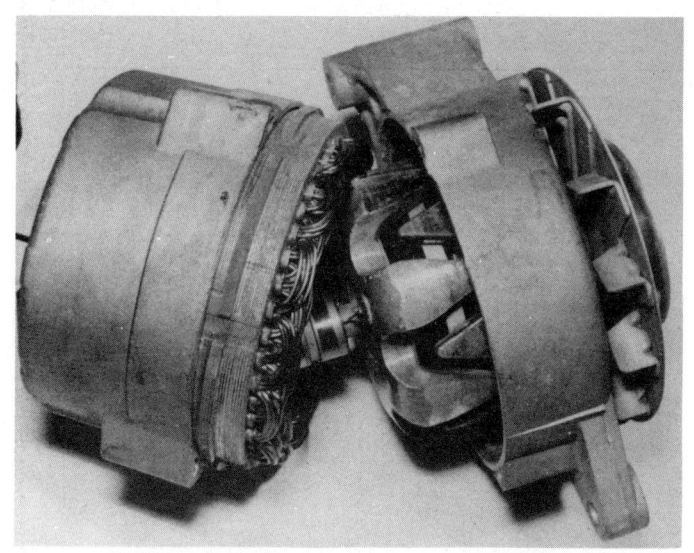

5.8 Separating the alternator housings

5.9 Alternator slip rings

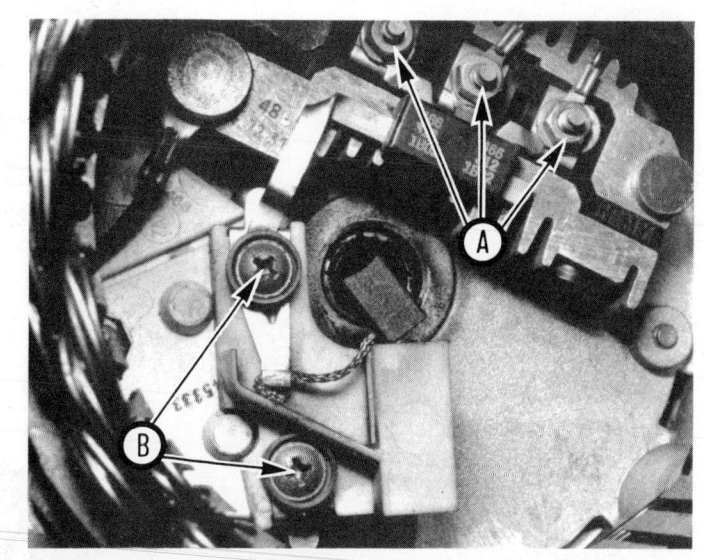

5.10 Stator lead retaining nuts (A) and brush holder securing screws (B)

5.14 Alternator brushes held retracted with a small twist drill (arrowed)

Chapter 7 Electrical system

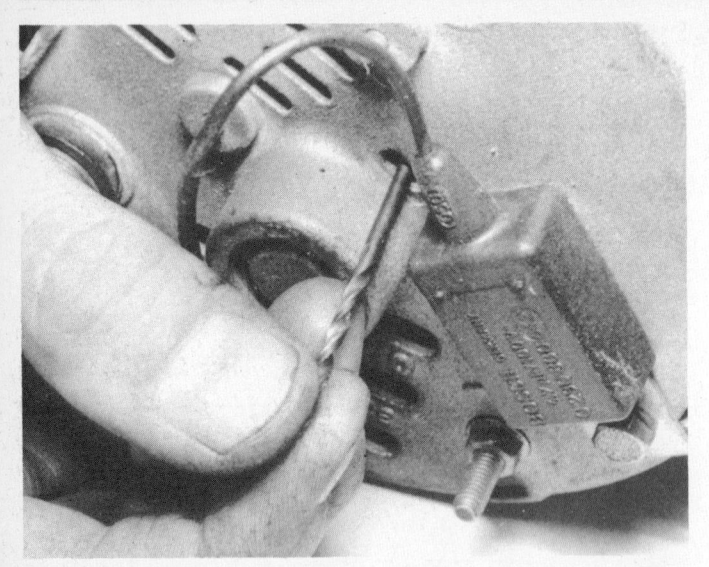

5.17 Withdraw the drill to release the brushes

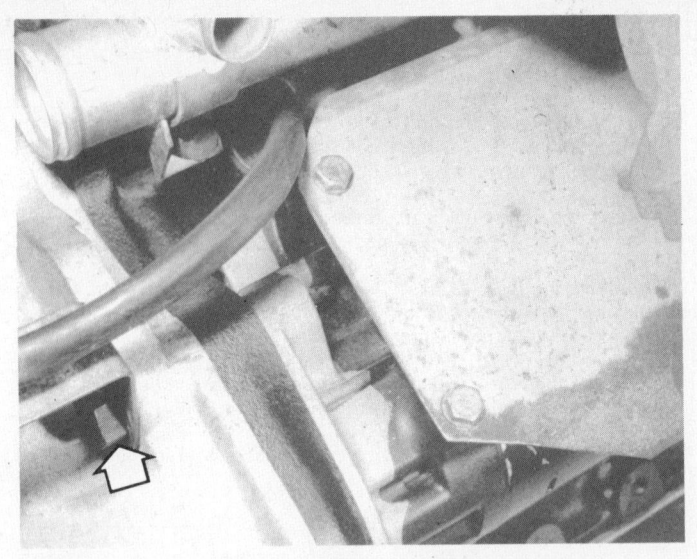

6.3 Starter motor upper retaining bolt (arrowed)

6 Starter motor – removal and refitting

Access to the starter motor in situ is very poor; it can be improved slightly by removing the exhaust downpipe, or a good deal by removing both manifolds.

1 Disconnect the battery earth lead.

2 Disconnect the starter motor supply lead from the battery positive terminal. Separate the command lead connector at the same point.

3 Remove the upper retaining bolt, which also secures a cable bracket (photo).

4 From below remove the lower retaining bolt, and the bolt(s) securing the tail bracket to the block (photo).

5 Remove the starter motor complete with heat shield, leads and tail bracket.

6 If a new starter motor is to be fitted, transfer the accessories to it. Pay attention to the supply lead, which must be fitted so that no metal part of it will touch the heat shield or block (photo).

7 Fit the assembled motor, with bracket, leads and heat shield, and secure it by nipping up the lower retaining bolt.

8 Fit the tail bracket-to-block bolt(s). If the bracket is under strain, slacken the bracket-to-starter nuts while tightening the bolt(s), then retighten them on completion.

9 Fit the upper retaining bolt and tighten it to the specified torque. Do not forget the cable bracket.

10 Tighten the lower retaining bolt to the specified torque.

11 Reconnect the starter motor leads and the battery earth lead.

7 Starter motor (reduction gear type) – dismantling and reassembly

Note: *The permanent magnets which are used in this motor instead of field coils are sensitive to impact and pressure. Do not drop the motor, therefore, nor grip its body in the vice.*

1 If not already done, remove the heat shield, tail bracket and leads.

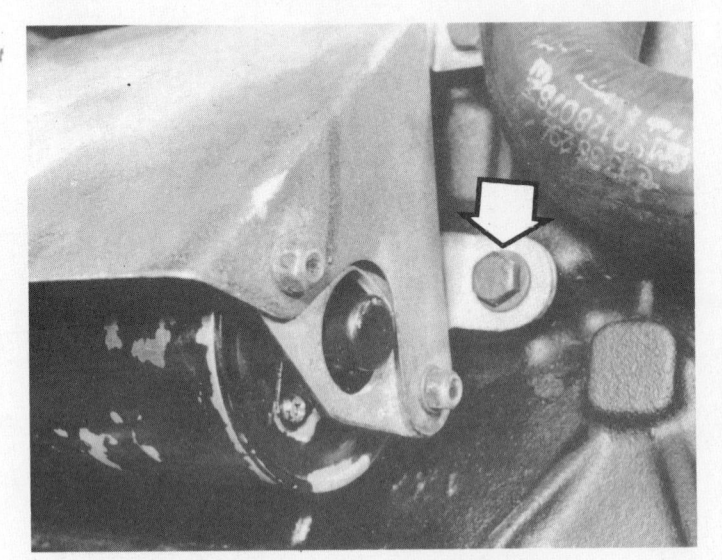

6.4 Starter motor tail bracket-to-block bolt (arrowed)

6.6 Starter motor with heat shield removed, showing correct fitting of supply lead terminal (arrowed)

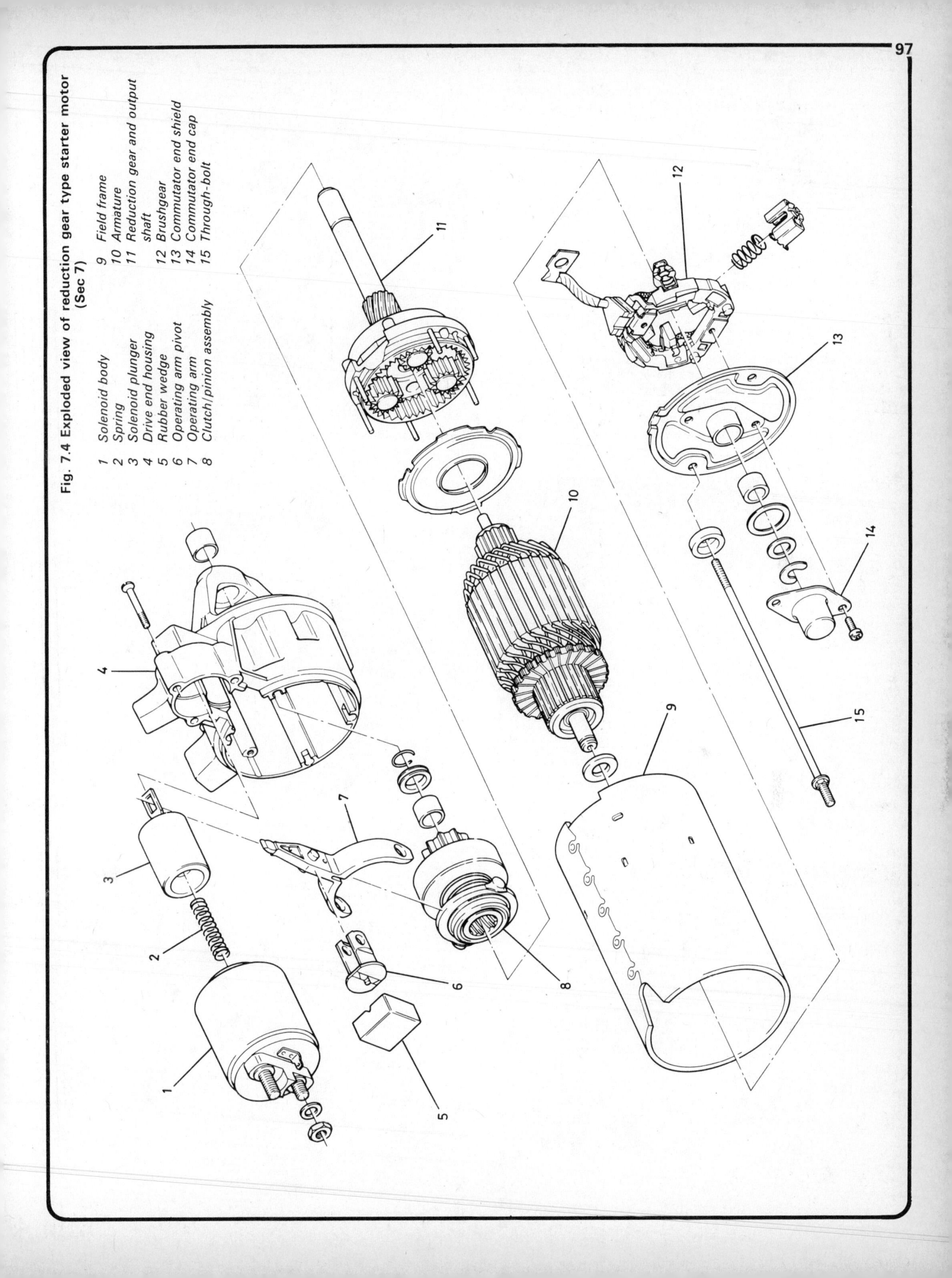

Fig. 7.4 Exploded view of reduction gear type starter motor (Sec 7)

1 Solenoid body
2 Spring
3 Solenoid plunger
4 Drive end housing
5 Rubber wedge
6 Operating arm pivot
7 Operating arm
8 Clutch/pinion assembly
9 Field frame
10 Armature
11 Reduction gear and output shaft
12 Brushgear
13 Commutator end shield
14 Commutator end cap
15 Through-bolt

Chapter 7 Electrical system

2 Remove the two screws which secure the commutator end cap. Remove the cap and sealing washer.

3 Remove the C-washer and plain washer(s) from the end of the armature shaft (photo).

4 Remove the two through-bolts from the commutator end shield. In the absence of the deep 7 mm box spanner needed to deal with these, lock two nuts onto the protruding threaded section of each bolt and unscrew them with a spanner on the nuts (photo).

5 Remove the commutator end shield from the field frame.

6 Disconnect the motor feed lead from the solenoid (photo).

7 Carefully withdraw the field frame, armature and brushgear from the solenoid and drive mechanism.

8 Undo the three screws which retain the solenoid. Unhook the solenoid from the operating lever and remove the plunger and spring (photos). Remove the drive mechanism from the drive end housing.

9 Remove the pinion/clutch assembly as follows. Drive the retaining ring down the shaft with a tube or socket to expose the snap-ring

7.3 Removing the C-washer (arrowed) from the end of the armature shaft

7.4 Undoing a through-bolt with two nuts locked together

7.6 Disconnecting the starter motor feed lead from the solenoid

7.8A Removing a solenoid retaining screw

7.8B Disengaging the solenoid from the operating lever

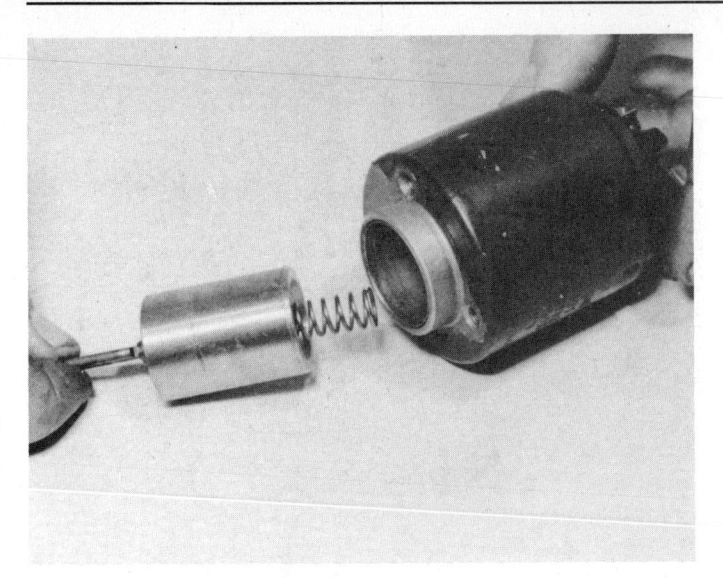

Chapter 7 Electrical system

7.8C Solenoid body, spring and plunger

Fig. 7.5 Removing the snap-ring from the output shaft (Sec 7)

(photo). Spread the ends of the snap-ring and remove it; obtain a new one for reassembly. Withdraw the pinion/clutch assembly from the reduction gear output shaft (photo).

10 The brushgear can be removed after prising out the grommet (photo). Be prepared for the sudden release of the brushes and springs.

11 There is now nothing holding the armature in place except the field magnets. As will be noticed, their force is quite strong.

12 The starter motor is now dismantled as far as is practicable. Apart from simple cleaning of the commutator, testing and repair of the armature should be left to an auto electrician. If many new components are needed, it may be cheaper and more satisfactory to fit a new or reconditioned motor. The reduction gears normally make a noise in use: this is not a fault.

13 Commence reassembly by lubricating the reduction gears with molybdenum disulphide paste or a similar 'dry' lubricant. Apply the same lubricant to the output shaft splines and to the solenoid lever pivot and rubbing points. Keep lubricant away from the commutator and brushgear.

14 Fit the pinion and clutch to the gear output shaft. Make sure that the retaining ring is in place, then fit a new snap-ring into the groove. Lever the ring upwards with a couple of spanners so that the snap-ring is compressed into its groove (photo).

15 Solder new brushes to the old brush leads if necessary, or fit new brushes complete with clips and leads. Make sure that all the springs are present and that the brushes are free to move in their holders (photos).

16 Prepare the brushgear for fitting by retracting the brushes with a tube of diameter equal to the commutator (photo) – a socket spanner is used in this instance.

17 Reassemble the solenoid and drive end components and insert them into the drive end housing. The gear carrier and its cover plate are located in the housing by pegs. Make sure that the rubber wedge is correctly fitted on the operating arm pivot.

18 Offer the armature and field frame to the drive end housing, engaging the armature shaft gear in the centre of the reduction gears (photo).

7.9A Driving the snap-ring retaining ring down the shaft

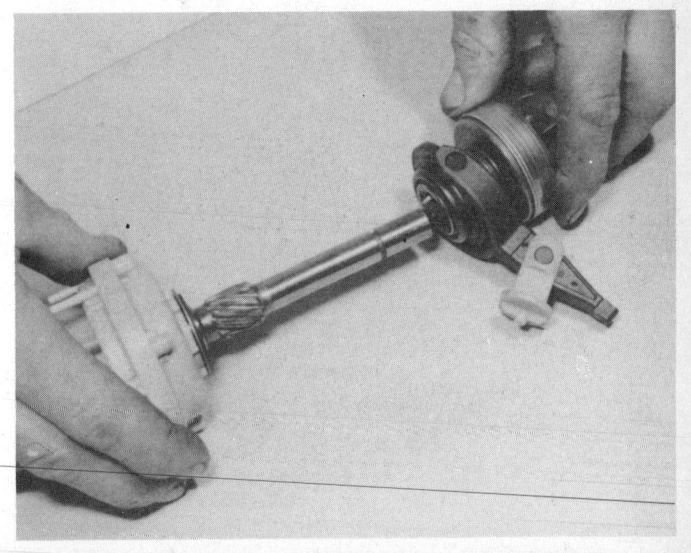

7.9B Removing the pinion/clutch assembly from the shaft

7.10 Prise out the grommet (arrowed) to remove the brushgear

7.14 Levering the retaining ring over the snap-ring

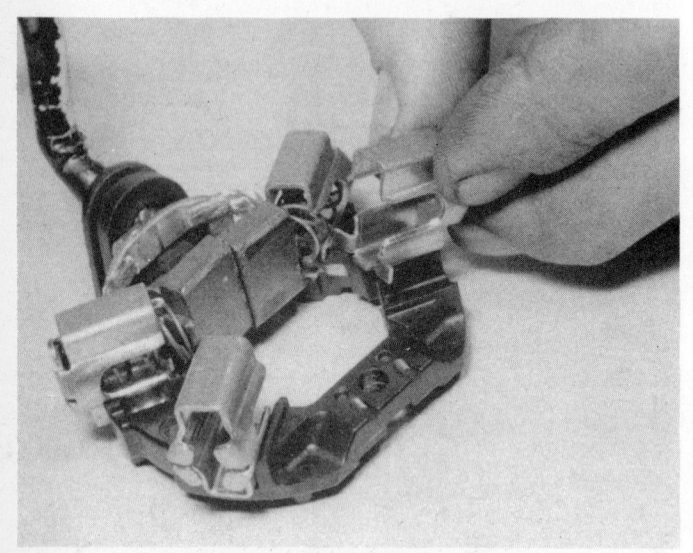

7.15A Fitting a brush holder to the carrier plate

7.15B Fitting a pair of brushes, with clips and leads, to the carrier plate

7.16 Keeping the brushes retracted with a tube – in this case a large socket

7.18 Fitting the armature and field frame to the drive end components

Chapter 7 Electrical system

19 Fit the brushgear over the commutator so that the tube is displaced and the brushes fall onto the commutator. Make sure that the feed lead grommet is correctly positioned.

20 Reconnect the feed lead to the solenoid.

21 Refit the commutator and shield and secure it with the through-bolts.

22 Refit the plain washer(s) and C-washer to the end of the armature shaft.

23 Press some grease into the commutator end cap, then fit the cap and its sealing washer (photo). Secure the cap with the two screws.

24 Refit the supply leads, tail bracket and heat shield.

8 Starter motor (direct drive type) – dismantling and reassembly

1 The direct drive starter motor is of the conventional series wound field type.

2 Remove the heat shield, tail bracket and supply leads (if not already done).

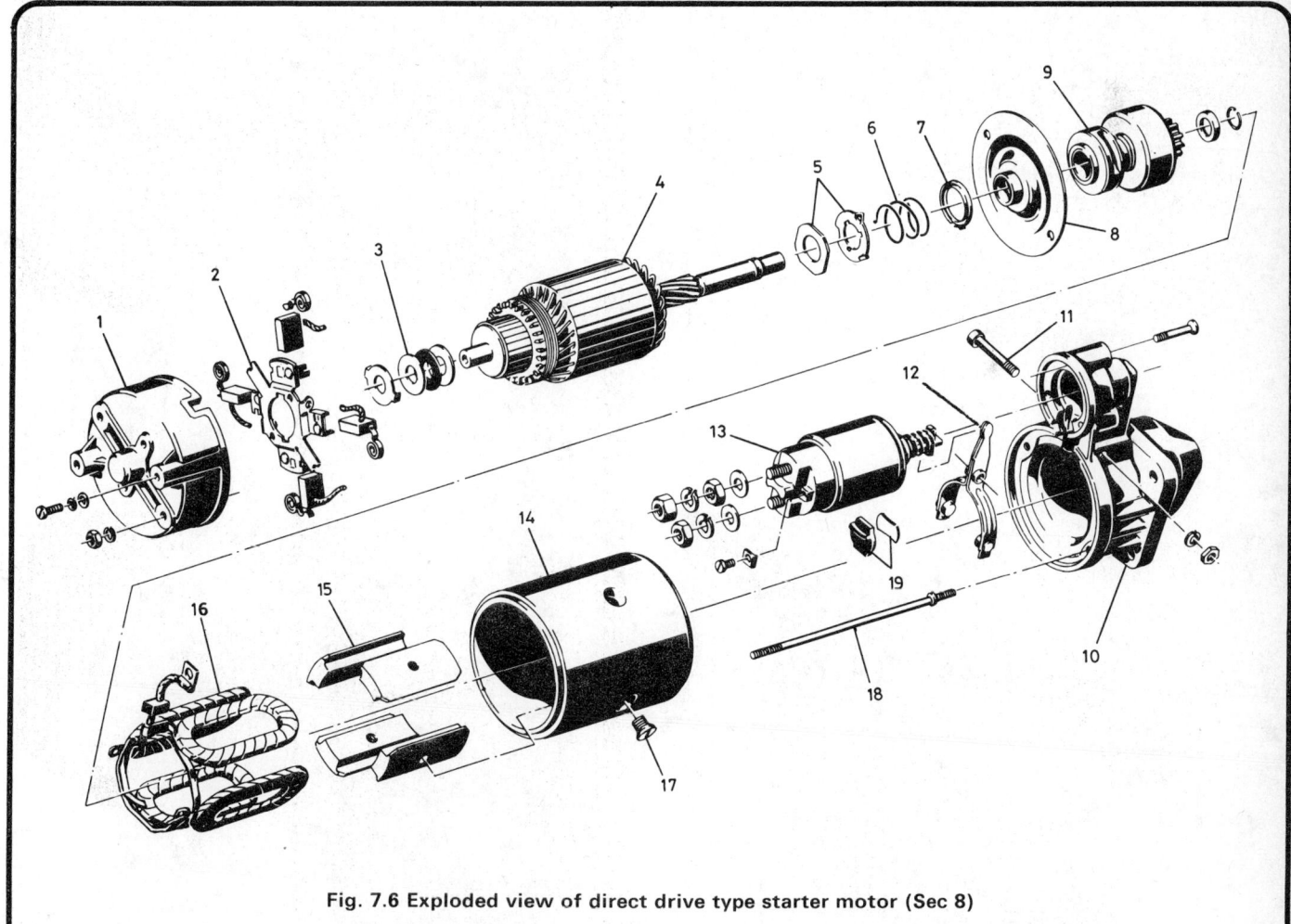

7.23 Fitting the commutator end cap and sealing washer

Fig. 7.6 Exploded view of direct drive type starter motor (Sec 8)

1 Commutator end cover
2 Brushgear
3 Shims and brake washers
4 Armature
5 Brake washers
6 Spring
7 Thrust washer
8 Intermediate bearing
9 Clutch/pinion assembly
10 Drive end housing
11 Operating arm pivot
12 Operating arm
13 Solenoid
14 Field frame
15 Pole pieces
16 Field coils
17 Pole piece screw
18 Through-bolt
19 Metal plug and rubber seal

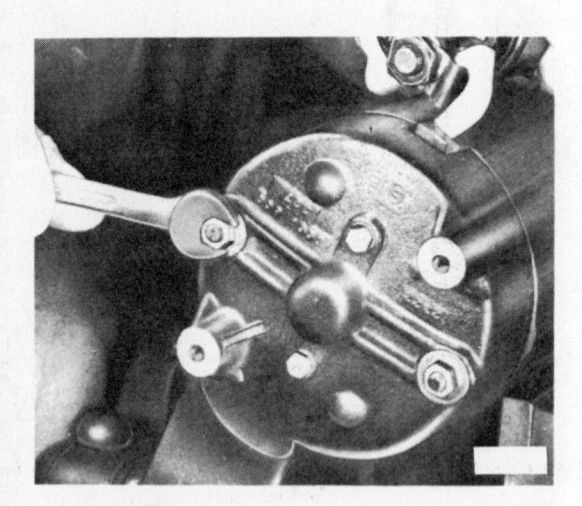

Fig. 7.7 Removing a commutator end cover nut (Sec 8)

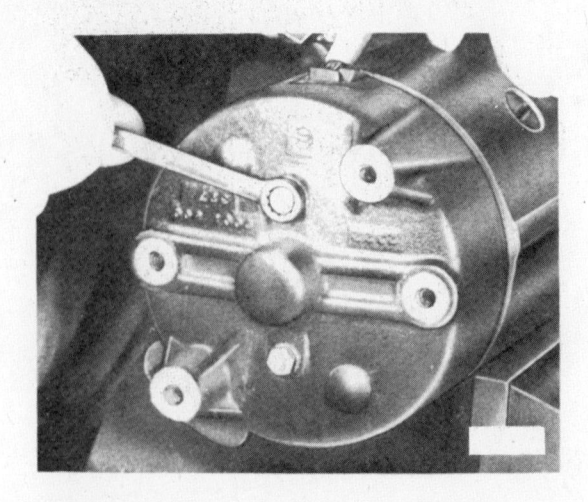

Fig. 7.8 Removing a brushgear mounting screw (Sec 8)

3 Remove the two nuts and washers from the commutator end cover.

4 Remove the two brushgear mounting screws from the commutator end cover. Carefully remove the cover.

5 Disconnect the motor feed lead from the solenoid.

6 Withdraw the field (positive) brushes from their holders, noting the arrangement of the springs. Remove the brushgear from the armature shaft; recover and note the position of any shims and brake washers.

7 Carefully withdraw the field frame, with field brushes, coils and feed lead, from the armature and drive end housing.

8 Remove the three screws which secure the solenoid. Also unscrew the solenoid operating arm pivot.

9 Withdraw the solenoid, operating arm, armature and drive components from the drive end housing. Note the location of the rubber seal and metal plug in the housing.

10 Remove the pinion and clutch from the armature shaft as described in Section 7, paragraph 9. Recover the intermediate bearing, thrust washer, spring and brake washers.

11 Clean and inspect all parts. Refer to Section 7, paragraph 12. Also check the field coils and the field (positive) brush holders for shorts to earth, using an ohmmeter or self-powered test lamp. Have the field winding renewed, or fit new brushgear, if shorts are found.

12 If it is necessary to renew the brushes only, cut off the old ones and solder the new brush leads to the old tails. Do not allow solder to run into the flexible parts of the leads near the brushes.

13 Lubricate the armature shaft splines, and the operating arm pivot and rubbing points, with molybdenum disulphide paste or a similar 'dry' lubricant. Apply a spot of clean engine oil to the armature bearings. Keep lubricant off the commutator and brushgear.

14 Refit the brake washers, spring, thrust washer and intermediate bearing to the armature shaft. Make sure that the bent end of the spring engages in the cut-out in the thrust washer.

15 Refit the clutch, pinion and retaining ring. Fit a new snap-ring into the groove and compress it by levering the retaining ring over it with a couple of spanners.

Fig. 7.9 Brushgear exposed by removal of end cover (Sec 8)

Fig. 7.10 Unscrewing the solenoid operating arm pivot (Sec 8)

Chapter 7 Electrical system 103

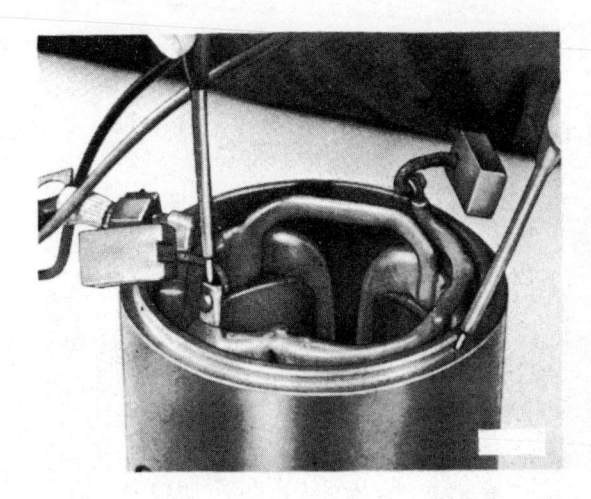

Fig. 7.11 Checking the field coils for a short to earth (Sec 8)

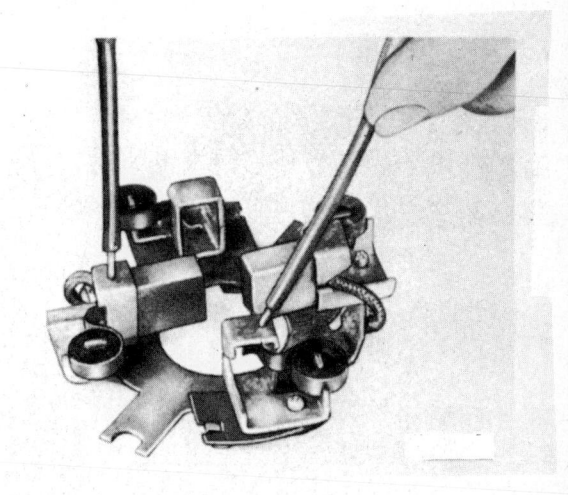

Fig. 7.12 Checking a positive brush holder for a short to earth (Sec 8)

16 Fit the armature and operating arm to the drive end housing. Refit the metal plug and the rubber seal, then engage and secure the solenoid and the operating arm pivot bolt.

17 Refit the field frame with coils, brushes etc.

18 Refit the brake washers, shims and brushgear. Make sure that the brushes move freely in their holders and that (when installed) the springs bear on the brushes.

19 Reconnect the feed lead to the solenoid.

20 Refit the commutator end cover, aligning the brushgear securing holes when doing so. Fit and tighten the two brushgear screws.

21 Refit the two washers and nuts which secure the end cover.

22 Refit the supply leads, tail bracket and heat shield.

9 Wiring diagrams – general

GM wiring diagrams are 'universal', so no separate diagrams are published for Diesel models. Refer to the appropriate manual for petrol-engined models.

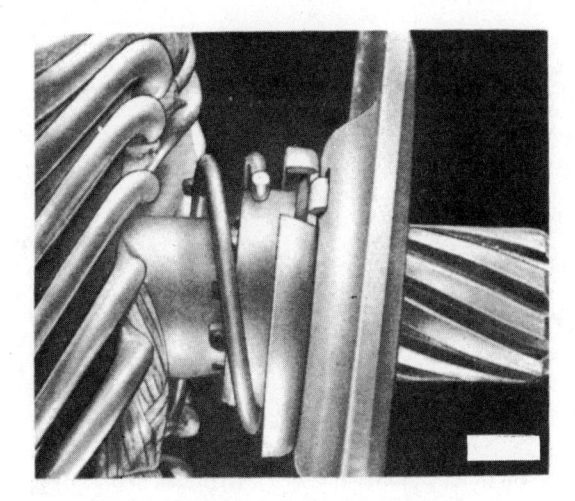

Fig. 7.13 Detail showing spring engagement with thrust washer (Sec 8)

Chapter 8 Supplement:
Revisions and information on later models

Contents

Introduction	1
Specifications	2
Routine maintenance	3
Maintenance schedule – 1992 Model Year onwards	
Engine	4
Introduction to the 17D and 17DR engines	
Camshaft drivebelt (17D engine) – inspection, removal, refitting and tensioning	
Camshaft drivebelt (17DR engine) – inspection, removal, refitting and tensioning	
Valve timing (16DA engines from May 1989 onwards, and all 17D and 17DR engines) – checking and adjustment	
Cylinder head (17D and 17DR engines) – removal and refitting	
Hydraulic valve lifters – inspection	
Cylinder bores – examination and renovation	
Engine/transmission mountings (17D and 17DR engines) – renewal	
Engine (17D and 17DR) – removal and refitting	
Cooling system	5
Water pump (17D and 17DR engines) – removal and refitting	
Fuel and exhaust systems	6
Maintenance and inspection	
Heated fuel filter	
Air cleaner – element renewal	
Air cleaner housing – removal and refitting	
Idle and maximum speed adjustment – Bosch VE fuel injection pump	
Idle and maximum speed adjustment – Lucas/CAV fuel injection pump	
Fuel injection pump timing (17D and 17DR engines) – checking and adjustment	
Vacuum-operated cold start device (17DR engines) – removal and refitting	
Preheating system – notes on testing	
Exhaust gas recirculation (EGR) system – 17DR engines	

1 Introduction

This Supplement details modifications to the Vauxhall/Opel Diesel engine used in the Astra, Belmont and Cavalier models, including the Astra-derived Bedford Vans, since this manual was first published. It contains information on the later 16DA engine unit, including certain procedural or specification revisions which apply retrospectively.

The bulk of the Supplement however, describes changes applicable to the 17D and 17DR engines introduced into the UK Astra/Belmont and Cavalier model ranges during the early part of 1989 and 1993 respectively. The information included is that which is additional to, or a revision of, material in the preceding Chapters of this manual.

It is recommended that before any operation is undertaken, reference is made to the appropriate Section(s) of this Supplement. In this way, any changes can be noted before reference is made to the main text.

2 Specifications

Engine – 1.6 litre (16DA)

Torque wrench settings

The torque wrench settings shown below are only those which have been revised or amended by the manufacturer since the original publication of this manual. Where no figure is quoted below, refer to Chapter 1 Specifications.

	Nm	lbf ft
Crankshaft pulley/sprocket centre bolt*:		
Stage 1	130	96
Stage 2	Angle-tighten a further 45°	
Flywheel to crankshaft*:		
Stage 1	50	37
Stage 2	Angle-tighten a further 30°	
Main bearing caps*:		
Stage 1	50	37
Stage 2	Angle-tighten a further 45°	
Connecting rod caps*:		
Stage 1	35	26
Stage 2	Angle-tighten a further 30°	

Chapter 8 Supplement: Revisions and information on later models

Torque wrench settings (continued) | **Nm** | **lbf ft**

Cylinder head bolts*:
- Stage 1 .... 25 | 18
- Stage 2 .... Angle-tighten a further 60°
- Stage 3 .... Angle-tighten a further 60°
- Stage 4 .... Angle-tighten a further 60°
- Stage 5 (after warm-up) .... Angle-tighten a further 30°
- Stage 6 (after 650 miles/1000km) .... Angle-tighten a further 45°

Glow plugs .... 20 | 15

*Bolts tightened by the angular method **must** be renewed every time.

Engine – 1.7 litre (17D and 17DR)
The following specifications relate to the 17D and 17DR engines only where they differ from their 16DA predecessor. Where no information appears here, refer to the 16DA specifications in the preceding Chapters.

General
- Bore .... 82.5 mm
- Displacement .... 1699 cc
- Maximum power:
  - 17D .... 42 kW (57 bhp) @ 4600 rpm
  - 17DR .... 44 kW (60 bhp) @ 4600 rpm
- Maximum torque .... 105 Nm (77.5 lbf ft) @ 2400 rpm

Cylinder head
- Sealing surface finish – peak-to-valley height .... 0.020 mm
- Deviation of sealing surface from true .... 0.05 mm max

Camshaft and bearings
Camshaft identification marks:
- Letter C
- Camshaft run-out .... 0.04 mm
- Camshaft endfloat .... 0.09 to 0.21 mm
- Cam lift (inlet and exhaust) .... 5.80 mm

Cylinder bores:	**Diameter***	**Identification**
Production grade 1	82.45 mm	5
	82.46 mm	6
	82.47 mm	7
Production grade 2	82.48 mm	8
	82.49 mm	99
	82.50 mm	00
	82.51 mm	01
	82.52 mm	02
	82.53 mm	03
Production grade 3	82.54 mm	04
	82.55 mm	05
	82.56 mm	06
	82.57 mm	07
	82.58 mm	08
	82.59 mm	09
	82.60 mm	1
Oversize (0.5 mm)	82.97 mm	7 + 0.5
	82.98 mm	8 + 0.5
	82.99 mm	9 + 0.5
	83.00 mm	0 + 0.5
Oversize (1.0 mm)	83.47 mm	7 + 1.0
	83.48 mm	8 + 1.0
	83.49 mm	9 + 1.0
	83.50 mm	0 + 1.0

*Tolerance ± 0.005 mm

Pistons
- Diameter .... 0.020 to 0.040 mm less than bore diameter

Piston rings
Oil scraper ring:
- Thickness .... 2.975 to 2.990 mm
- End gap (fitted) .... 0.250 to 0.500 mm

Connecting rods
- Weight variation (in same engine) .... 8 grammes max

Chapter 8 Supplement: Revisions and information on later models

Lubrication system
Lubricant type/specification .... Multigrade engine oil, viscosity range SAE 10W/40 to 20W/50, to API SG/CD or better (Duckhams Diesel, QS, QXR, Hypergrade Plus, or Hypergrade)
Lubricant capacity (drain and refill, including filter) .... 4.75 litres (8.4 pints) approx
Oil pressure (at idle, engine warm) .... 1.0 bar (14.5 lbf/in^2)

Torque wrench settings
The following torque wrench settings for the 17D and 17DR engines are only those which differ from their 16DA predecessor. Where no information appears here, refer to the 16DA settings given at the start of these Specifications, or to Chapter 1.

	Nm	lbf ft
Crankshaft pulley/sprocket centre bolt*:		
Stage 1	145	107
Stage 2	Angle-tighten a further 30°	
Stage 3	Angle-tighten a further 10°	
Main bearing caps*:		
Stage 1	50	37
Stage 2	Angle-tighten a further 45°	
Stage 3	Angle-tighten a further 15°	
Connecting rod caps*:		
Stage 1	35	26
Stage 2	Angle-tighten a further 45°	
Stage 3	Angle-tighten a further 15°	
Cylinder head bolts*:		
Stage 1	25	18
Stage 2	Angle-tighten a further 90°	
Stage 3	Angle-tighten a further 90°	
Stage 4	Angle-tighten a further 45°	
Warm-up engine, then:		
Stage 5	Angle-tighten a further 30°	
Stage 6	Angle-tighten a further 15°	
Camshaft sprocket bolt*:		
Stage 1	75	55
Stage 2	Angle-tighten a further 60°	
Sump bolts**	5	4
Starter motor to block:		
Engine side	45	33
Transmission side	75	55
Oil pressure switch	30	22
Rear engine mounting bracket to transmission	60	44
Right-hand engine mounting bracket to cylinder block	60	44
Left-hand engine mounting to bracket	60	44
Right-hand engine mounting to bracket	35	26
Rear engine mounting to bracket	45	33
Rear engine mounting to crossmember	40	30
Right and left-hand engine mountings to side member*	65	48

*Bolts **must** be renewed every time
**Use thread-locking compound

Fuel and exhaust systems – 17D and 17DR engines
General
Fuel filter type:
  Cavalier from 1988, and Astra to 1991 .... Champion L113
  Astra from 1991, and Astramax .... Champion L111
Air filter type:
  Astra 1988 to 1991, and Astramax .... Champion U558
  Astra from 1991 .... Champion U599
  Cavalier from 1988 .... Champion U554

Injection pump
Type .... Bosch or Lucas/CAV
Maker's identification:
  Bosch .... VE 4/9F 2300 R 313, 443 or 487
  Lucas/CAV .... OP 02 DPC R8443 B55 OA or OP 03 DPC R8443 B85 OC

Injectors
Maker's identification:
  Bosch .... DN OSD 309
  Lucas/CAV .... BDN OSD C 6751 D or RDN OSD C 6751 D
Opening pressure (Bosch and Lucas/CAV):
  New .... 135 to 143 bars (1958 to 2073.5 lbf/in^2)
  Used .... 130 to 138 bars (1885 to 2001 lbf/in^2)

Glow plugs
Rating .... 5 volts

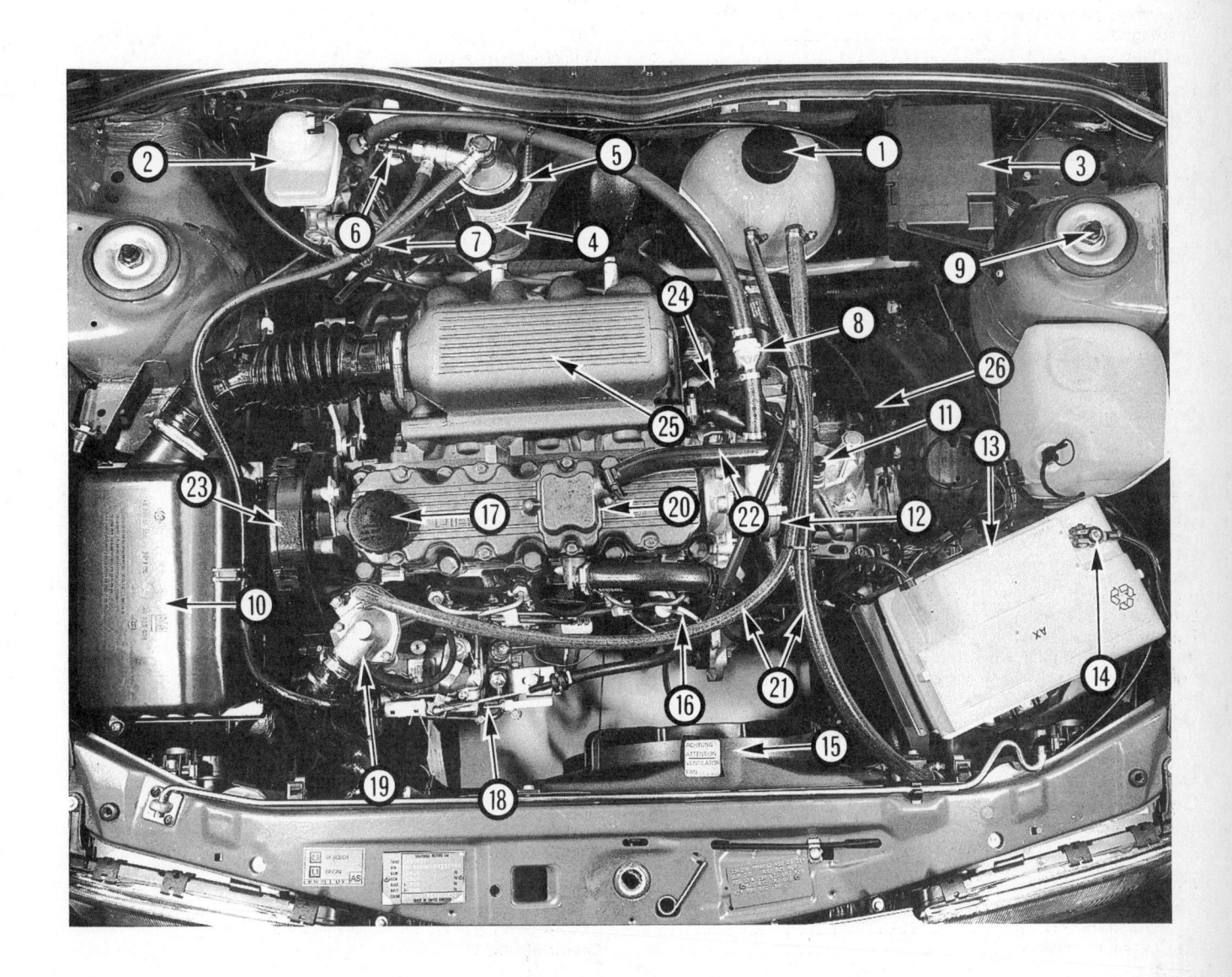

Chapter 8 Supplement: Revisions and information on later models   107

Under-bonnet view of a typical 17D engine

1 Cooling system filler/pressure cap	7 Fuel hoses	14 Battery earth strap	21 Cooling system vent hoses
2 Brake fluid reservoir cap	8 Brake servo non-return valve	15 Radiator fan	22 Crankcase ventilation hose
3 Electrical ancillary box	9 Suspension turret	16 Engine oil dipstick	23 Camshaft drivebelt cover
4 Fuel filter	10 Air cleaner housing	17 Engine oil filler	24 Coolant hose (to expansion tank)
5 Fuel filter heater unit	11 Gearbox breather	18 Fuel injection pump	25 Inlet manifold
6 Fuel filter temperature sensor	12 Vacuum pump	19 Thermostat elbow	26 Clutch cable
	13 Battery	20 Engine breather	

Chapter 8 Supplement: Revisions and information on later models

Adjustment data

Idle speed:	
17D engine	820 to 920 rpm
17DR engine (see text, Section 6):	
Engine temperature below 20°C (68°F)	1200 rpm
Engine temperature above 20°C (68°F)	820 to 920 rpm
Governed maximum speed	5500 to 5600 rpm
Injection commencement at idle	2° to 4° BTDC
Pump timing setting (see text, Section 6):	
Bosch injection pump:	
17D engine	0.80 + 0.05 mm (0.031 + 0.0019 in)
17DR engine	0.85 + 0.05 mm (0.033 + 0.0019 in)
Lucas/CAV injection pump	X – 0.15 mm (0.005 in) where X = maker's calibration marked on pump

Torque wrench setting

	Nm	lbf ft
Glow plugs	20	15

Suspension and steering

Tyre pressures (cold) – bars (lbf/in^2)

	Front	Rear
Astra from 1991:		
Light load	2.0 (29)	1.7 (24.5)
Full load	2.2 (32)	2.4 (35)
Cavalier from 1988:		
Light load	1.9 (27)	1.7 (24)
Full load	2.1 (30)	2.3 (33)

3 Routine maintenance

Maintenance schedule – 1992 Model Year onwards

General

1 A revised maintenance schedule has been introduced for the 1992 model year. As before, it is based on the assumption of an average annual mileage of 9000 miles (15 000 km); the service to be carried out annually or at this mileage, whichever occurs first. However, the new schedule includes a high-mileage inspection for vehicles covering more than 18 000 miles (30 000 km) per year, and also makes provision for vehicles which cover very low mileages.

2 Essential points to note are as follows:

(a) The engine oil must be changed, and the oil filter renewed, at least annually or every 4500 miles (7500 km), whichever occurs first.

(b) The remaining thickness of friction material on the disc brake pads, and on the rear brake shoes, must be checked at the interval given at the front of this manual. If any pad or shoe is excessively worn when checked, or is likely to be worn out before the next service interval, all four pads/shoes must be renewed (ie, as an axle set).

(c) Ensure that the brake fluid is renewed annually, regardless of mileage.

(d) 17DR engines are fitted with an automatic tensioner for the camshaft drivebelt; on these engines, the manufacturer no longer calls for routine checking of the drivebelt. However, due to the importance of the timing belt to the engine's reliability, owners are advised to check the belt's condition (noting the differences mentioned in Section 4 of this Chapter), at the interval given in the Routine maintenance at the front of this manual. If the belt is found to be worn, damaged or in any way suspect, it must be renewed as soon as possible. Obviously, the belt's tension, once set on installation, does **not** require checking or adjustment.

Basic service every 1st, 3rd, 5th, 7th, etc, interval

3 This service is to be carried out every 12 months or 9000 miles/ 15 000 km, whichever occurs first, and every **other** interval after that. Carry out the following:

(a)* Change the engine oil and renew the oil filter (Chapter 1, Section 4). This should be carried out every 4500 miles/7500 km, or annually, as noted in paragraph 2a above.

(b) Check the engine and transmission for leaks.

(c) Check the coolant level; top-up if necessary, and check the cooling system for leaks.

(d)* Drain any accumulation of water from the fuel filter (Chapter 3 and Section 6 of this Chapter).

(e)* Renew the air cleaner filter element (Section 6 of this Chapter).

(f) Check the engine idle speed, governed maximum speed and exhaust emissions (Section 6 of this Chapter). Note that the exhaust emissions test includes an inspection of pollution-relevant components (including the exhaust gas recirculation system, where fitted) and should be carried out by a Vauxhall/Opel dealer.

(g)* Check the remaining thickness of friction material on the disc brake pads; renew all four as an axle set if necessary.

(h) Check the operation of the load-dependent brake proportioning or regulating valve (Estate/Van models only, where applicable).

(i) Check the condition of the braking system hoses and pipes.

(j) Renew the brake fluid.

(k)* Check the condition and tension (where applicable) of the accessory drivebelt(s) (Chapter 7). The power steering drivebelt (where applicable) should be checked as described in the relevant petrol-engine manual.

(l)* Check the tightening torque of the roadwheel bolts. Check the tyres for wear, and check the tyre pressures (including the spare). If uneven tread wear is apparent, this may be due to wheel misalignment – have the alignment checked.

(m) Check the bodywork for damage, and the underbody for signs of damage to its protective coating, or of actual corrosion.

(n)* Check the operation of all interior and exterior lights, direction indicators/hazard warning lights, stop-lights, headlamp flasher, and the horn.

(o) Have the headlamp alignment checked.

(p) Check the operation of the windscreen (and, where applicable, the rear window and headlight) wipers and washers; renew the blades if worn, and top-up the washer fluid reservoir.

(q)* Road-test the vehicle, checking for correct operation of all controls, warning lights, switches and accessories.

Note: *All items above marked with an asterisk (*) are to be repeated at the additional high-mileage inspection (see above) for vehicles covering 18 000 miles (30 000 km) annually, after the first 9000 miles (15 000 km) inspection.*

Chapter 8 Supplement: Revisions and information on later models

Full service every 2nd, 4th, 6th, 8th, etc, interval

4 This service is to be carried out every 24 months or 18 000 miles/30 000 km, whichever is the sooner, and every **other** interval after that. In addition to all those tasks listed for the basic service, carry out the following:

(a) Renew the fuel filter (Section 6 of this Chapter).
(b) On 17D engines only, at every 4th interval (ie, every 48 months or 36 000 miles/60 000 km, whichever is the sooner) check the condition of the camshaft drivebelt, and renew it if necessary. If the belt is fit for further use, check its tension and adjust if required (Section 4 of this Chapter).
(c) On 17D engines only, at every 8th interval (ie, every 8 years or 72 000 miles/120 000 km, whichever is the sooner) renew the camshaft drivebelt regardless of condition (Section 4 of this Chapter).
(d) Check the clutch pedal travel, and adjust if necessary.
(e) Check the manual gearbox oil level, and top-up if necessary.
(f) Check the automatic transmission fluid level, and top-up if necessary.
(g) Check the rubber gaiters or bellows on the driveshafts, steering and suspension components for splits, lubricant leakage, wear or damage.
(h) Check the remaining thickness of friction material on the rear brake shoes, renewing all four as an axle set if necessary.
(i) Where applicable, check the power steering fluid level, and top-up if necessary.
(j) Check the operation of all door, bonnet and boot lid/tailgate locks and hinges; lubricate all moving parts as necessary.

4 Engine

Introduction to the 17D and 17DR engines

1 The 17D engine closely resembles the 16DA unit from which it is derived and which it replaces, but embodies a number of detail changes designed to improve flexibility, and to reduce noise levels and vibration.

2 The most significant revision is the increase in cylinder bore size from 80.0 mm to 82.5 mm, increasing the capacity and giving improved torque output, with a consequent improvement in mid-range acceleration. Internally, the unit is little changed from its predecessor, although the pistons have been lightened to reduce engine noise and vibration, and the cylinder head has received detail modifications. These include thickening, and reinforcement by means of internal ribbing, of the lower section, to reduce valve leakage at cranking speed, and thus improve cold starting.

3 External changes include a revised intake system, employing a new cast aluminium alloy manifold and resonator assembly, linked by flexible trunking to a remote filter unit located to the rear of the right-hand headlight. The camshaft cover has also been modified, with the oil filler now located at the right-hand end, and changes have been made to the camshaft drivebelt covers; two versions have been fitted since the introduction of this engine type.

4 In addition to the mechanical revisions to the engine, there have been a number of changes to the engine ancillaries. A new Lucas/CAV injection pump has been introduced, featuring an automatic cold start system in place of the previous manual arrangement. The new pump also includes an electronic speed governor, and a solenoid-operated fuel shut-off valve which allows the engine to be stopped by turning the key to the off position. Detail changes have also been made to the existing Bosch injection pump, which continues as an alternative standard fitment on these engines. An electrically-heated fuel filter is used, allowing the engine to operate at lower ambient temperatures without the risk of fuel waxing.

5 The 17DR engine is a further development of the 17D unit, and features revisions to the swirl chambers, resulting in an increased power output. An exhaust gas recirculation system is also employed, to ensure that the engine will meet existing and proposed diesel exhaust emission regulations (see Section 6).

6 The engine is also fitted with a spring-loaded automatic camshaft drivebelt tensioner, to ensure correct belt tensioning on assembly, and to eliminate the need for regular belt retensioning as part of the routine maintenance schedule.

Camshaft drivebelt (17D engine) – inspection, removal, refitting and tensioning

7 The camshaft drivebelt arrangement employed on the 17D version of the engine is largely unchanged from its predecessors, and as such can be dealt with as described in Chapter 1, Section 5. Note, however, that revisions to the belt covers and air filter assembly dictate some changes of approach when working on the drivebelt components.

8 Both the Cavalier and the Astra/Belmont models have a new air filter system, with a remote filter housing attached to the inner wing area at the right-hand side of the engine bay. To gain access to the drivebelt covers, it will be necessary to remove the filter housing as described in Section 6 of this Chapter.

9 Two versions of the moulded plastic drivebelt covers have been used since the introduction of the 17D engine; the later type being identified by the squared-off top surface of the outer belt cover. On the earlier version, a screwdriver blade can be used to release the outer cover retaining clips, and the cover sections can then be removed as required to gain access to the drivebelt. On the later version, the method of retention is by bolts instead of clips. This arrangement is, in fact, the same as used on the 17DR engine, described in paragraphs 11 to 15 of the following section.

Camshaft drivebelt (17DR engine) – inspection, removal, refitting and tensioning

General

10 A spring-loaded automatic camshaft drivebelt tensioner is fitted to the 17DR engine. The tensioner automatically sets the drivebelt to the correct tension on assembly, and maintains that tension for the life of the drivebelt. The revised procedures for removal, inspection and refitting of the drivebelt, tensioner and related components on the 17DR engine are as follows.

Drivebelt outer covers – removal and refitting

11 The covers are a two-piece assembly, and it is necessary to remove the upper cover prior to removal of the lower cover. Remove the accessory drivebelt(s) (Chapter 7) and the air cleaner housing assembly (Section 6 of this Chapter) for access to the drivebelt covers. The power steering drivebelt (where applicable) should be removed as described in the relevant petrol-engine manual.

12 To remove the upper cover, undo the five securing bolts and lift off the cover (photo).

4.12 Removing the camshaft drivebelt upper cover

Chapter 8 Supplement: Revisions and information on later models

Fig. 8.1 Camshaft drivebelt automatic tensioner details – 17DR engine (Sec 4)

- A Alignment lugs on water pump and cylinder block
- B Tensioner pointer aligned with notch in tensioner bracket
- 1 Move the tensioner arm anti-clockwise to release the belt tension
- 2 Move the tensioner arm clockwise to tension the belt

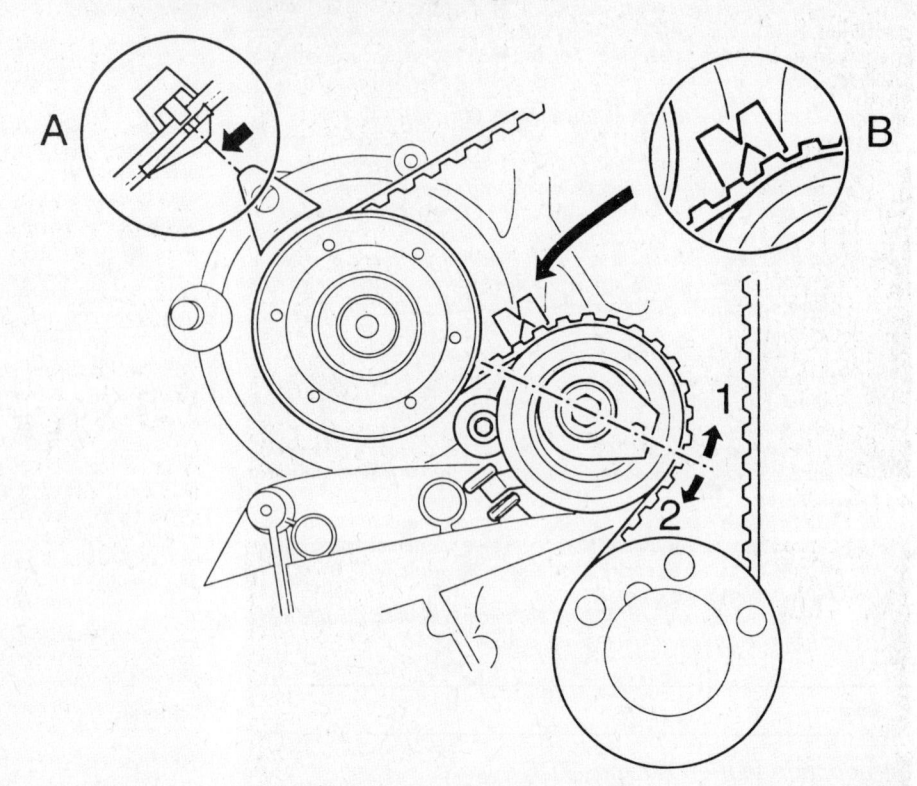

13   From under the front wheel arch, undo the four crankshaft pulley retaining bolts, and withdraw the pulley from the drivebelt sprocket (photo).

14   Undo the remaining three bolts, and remove the lower cover from the engine (photo).

15   Refitting the covers is a reversal of removal.

Drivebelt – removal, refitting and adjustment

16   The procedure for 17DR engines with an automatic drivebelt tensioner is essentially the same as described in Chapter 1, Section 5, except that it is **not** necessary to remove the engine mounting, nor to slacken the water pump mounting bolts and move the pump to adjust the belt tension. Instead, belt adjustment is catered for by means of the automatic tensioner, as follows.

17   To release the belt tension prior to removal, unscrew the drivebelt tensioner securing bolt slightly then, with a suitable Allen key inserted in the slot on the tensioner arm, turn the tensioner arm until the timing belt is slack (photo). Tighten the securing bolt slightly, to hold the tensioner in this position. The drivebelt can now be removed. Inspect the belt as described in Chapter 1, Section 5.

18   Prior to refitting the drivebelt, first ensure that the water pump is correctly positioned, by checking that the lug on the water pump flange is aligned with the corresponding lug on the cylinder block. If this is not the case, slacken the pump mounting bolts slightly and move the pump accordingly. Tighten the bolts to the specified torque on completion (see Chapter 2).

19   Initially refit the drivebelt as described in Chapter 1, Section 5, ensuring that No 1 piston is still at TDC, that the injection pump sprocket mark is still aligned and the camshaft position is still correct, then tension it as follows.

20   Slacken the automatic tensioner securing bolt, and move the tensioner arm anti-clockwise, until the tensioner pointer is at its stop. Tighten the tensioner securing bolt to hold the tensioner in this position.

21   Turn the crankshaft through two complete revolutions in the normal direction of rotation, until No 1 piston is once again at the TDC position. Check that the injection pump sprocket and camshaft sprocket positions are still correct.

22   Slacken the automatic tensioner securing bolt once again, and move the tensioner arm until the tensioner pointer and tensioner bracket notch coincide (Fig. 8.1). Tighten the tensioner securing bolt securely.

23   Check the valve timing as described below, and the injection pump timing (Section 6). Refitting the remainder of the components is the reversal of removal.

4.13 Crankshaft pulley retaining bolts

Chapter 8 Supplement: Revisions and information on later models

Valve timing (16DA engines from May 1989 onwards, and all 17D and 17DR engines) – checking and adjustment

24 The procedure described in Chapter 1, Section 6 can be applied to all engines up to May 1989. Subsequent 16DA, 17D and 17DR units are adjusted using the dial test indicator method described in the latter part of Section 6, but modified as described below. Before starting work, it is necessary to make sure that the drivebelt tension is set correctly.

25 You will need a dial test indicator (DTI) with a 10 mm (0.4 in) diameter measuring foot. Special tool KM-661-1 should ideally be available; this is a support bar that rests on the top face of the camshaft carrier, and positions the DTI above the camshaft. A home-made equivalent (such as that described in Chapter 1) can be used if tool KM-661-1 is not available.

26 An additional tool (KM-661-2) is prescribed by the manufacturer. This comprises a slotted steel plate with a stop screw which is secured by bolts to the camshaft carrier, immediately above the flats on the camshaft. The second part of the tool is effectively an open-jawed spanner, which fits over the flats on the camshaft, and passes up through the slotted plate. The stop screw bears on the spanner handle, allowing precise positioning of the camshaft. In the absence of the manufacturer's tool, it should not prove difficult to make up an equivalent device at home. Both these tools are depicted in the accompanying illustrations.

27 To check the valve timing, turn the crankshaft in the normal direction of rotation, and stop when the crankshaft is approximately 90° BTDC, with No 1 cylinder on the compression stroke. Fit the DTI to the support bar, and position the foot of the gauge over the base circle of the second cam from the sprocket end (No 1 cylinder inlet cam). Set the DTI to zero.

28 Carefully move the DTI and the support bar (without disturbing the position of the DTI in the support bar) exactly 10 mm (0.4 in) to the left, as viewed from the camshaft sprocket end of the engine (ie towards the peak of the cam lobe). Turn the crankshaft to the TDC position for No 1 cylinder (see Chapter 1, Section 5). In this position, the DTI should show a lift of 0.55 ± 0.03 mm (0.022 ± 0.001 in). If so, the valve timing is correct.

29 If adjustment is necessary, slacken the camshaft sprocket bolt, noting that since this must be renewed each time it is disturbed, it is as well to fit a new bolt loosely at this stage. Release the taper between the sprocket and the camshaft, if necessary by tapping the sprocket with a wooden or plastic mallet.

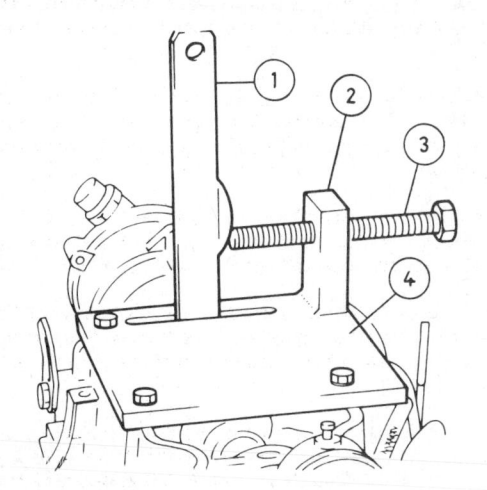

4.14 Removing the camshaft drivebelt lower cover

4.17 Release the drivebelt tensioner securing bolt, then turn the tensioner arm with an Allen key until the belt is slack

Fig. 8.2 Service tool KM-661-1 in use (Sec 4)

1 Open-jawed spanner
2 Stop screw bracket welded to baseplate
3 Stop screw
4 Baseplate located by camshaft cover bolts

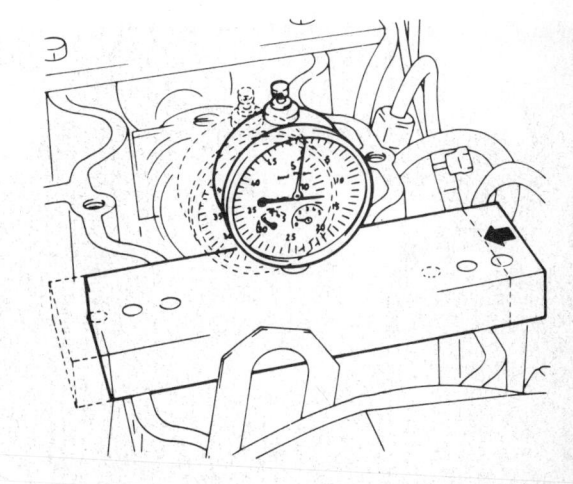

Fig. 8.3 Dial Test Indicator (DTI) in position above camshaft (Sec 4)

Note dotted lines indicating the two base positions during the test procedure

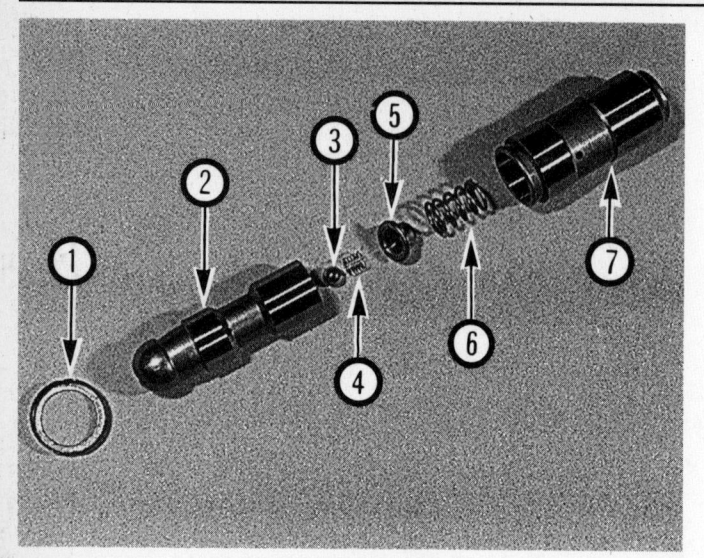

Chapter 8 Supplement: Revisions and information on later models

4.36 Hydraulic valve lifter components

1 Collar
2 Plunger
3 Ball
4 Small spring
5 Plunger cap
6 Large spring
7 Cylinder

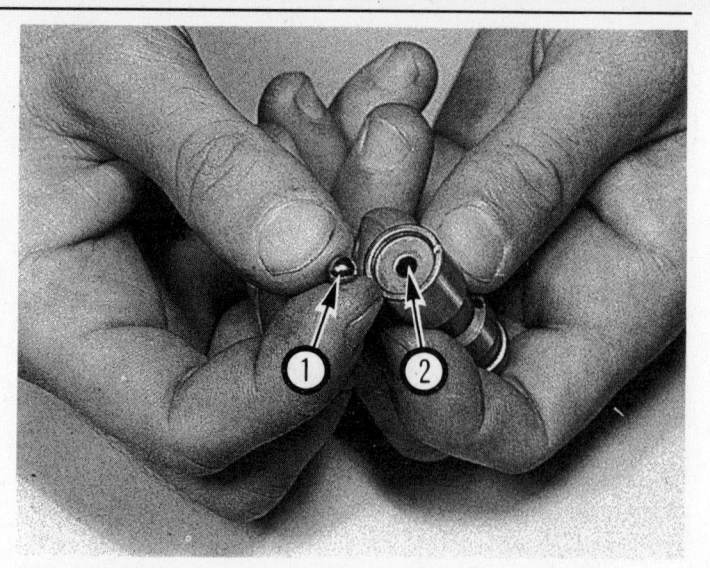

4.43 Locate the ball (1) on its seat (2) in the base of the plunger

30 Using the flats on the camshaft, turn it until the DTI reads approximately 0.80 mm (0.031 in) of lift. Check that the crankshaft is still set to TDC.

31 Assemble and fit the holding tool, KM-661-2 or equivalent. Using the stop screw, gradually set the cam lift to 0.60 to 0.64 mm (0.023 to 0.025 in). Tighten the camshaft sprocket bolt tight enough for the camshaft taper to lock the sprocket, then remove the holding tool.

32 Carefully lift away the DTI and its support bar, taking care not to disturb the DTI position in the bar. Turn the crankshaft through two complete revolutions, then position the DTI once more, and check that a lift figure of 0.55 ± 0.03 mm (0.021 ± 0.001 in) is shown at TDC. If the correct figure is not shown, repeat the adjustment sequence. If the figure is correct, tighten the (new) camshaft sprocket bolt to the specified torque, check the valve timing once more, then remove the tools.

33 Remember that the injection pump timing must be checked after any change in the valve timing setting. Refit the various covers removed during the checking operation.

Cylinder head (17D and 17DR engines) – removal and refitting

34 As mentioned in the introduction to this Section, certain modifications have taken place to the camshaft drivebelt covers and ancillary components on these engines, which slightly alter the procedures for certain operations.

35 The removal and refitting procedures contained in Chapter 1, Section 7 are essentially the same for the 17D and 17DR engines. However, note that it will be necessary to unbolt the camshaft drivebelt inner cover from its cylinder head attachments prior to removal. Also bear in mind that there are revisions to the intake manifold on both engines, and on the 17DR engine, exhaust gas recirculation system components will be encountered. Further details of these will be found in Section 6 of this Chapter.

Hydraulic valve lifters – inspection

36 On engines which have covered a high mileage, or for which the service history (particularly oil changes) is suspect, it is possible for the valve lifters to suffer internal contamination, which in extreme cases may result in increased engine top end noise and wear. To minimise the possibility of problems occurring later in the life of the engine, it is advisable to dismantle and clean the hydraulic valve lifters as follows whenever the cylinder head is overhauled. Note that no spare parts are available for the valve lifters, and if any of the components are unserviceable, the complete assembly must be renewed (photo).

37 With the cylinder head removed and dismantled as described in Chapter 1, first inspect the valve lifter bores in the cylinder head for wear. If excessive wear is evident, the cylinder head must be renewed. Also check the valve lifter oil holes in the cylinder head for obstructions.

38 Starting with No 1 valve lifter, carefully pull the collar from the top of the valve lifter cylinder. It should be possible to remove the collar by hand – if a tool is used, take care not to distort the collar.

39 Withdraw the plunger from the cylinder, and recover the spring.

40 Using a small screwdriver, carefully prise the cap from the base of the plunger. Recover the spring and ball from under the cap, taking care not to lose them as the cap is removed.

41 Carefully clean all the components using paraffin or a suitable solvent, paying particular attention to the machined surfaces of the cylinder (internal surfaces), and piston (external surfaces). Thoroughly dry all the components using a lint-free cloth. Carefully examine the springs for damage or distortion – the complete valve lifter must be renewed if the springs are not in perfect condition.

42 Lubricate the components sparingly with clean engine oil of the correct grade (see Chapter 1), then reassemble as follows.

43 Invert the plunger, and locate the ball on its seat in the base of the plunger (photo).

44 Locate the smaller spring on its seat in the plunger cap, then carefully refit the cap and spring, ensuring that the spring locates on the ball. Carefully press around the flange of the cap, using a small screwdriver if necessary, until the flange is securely located in the groove in the base of the plunger (photos).

45 Locate the larger spring over the plunger cap, ensuring that the spring is correctly seated, and slide the plunger and spring assembly into the cylinder (photos).

46 Slide the collar over the top of the plunger, and carefully compress the plunger by hand, until the collar can be pushed down to engage securely with the groove in the cylinder (photo).

47 Repeat the above procedures on the remaining valve lifters.

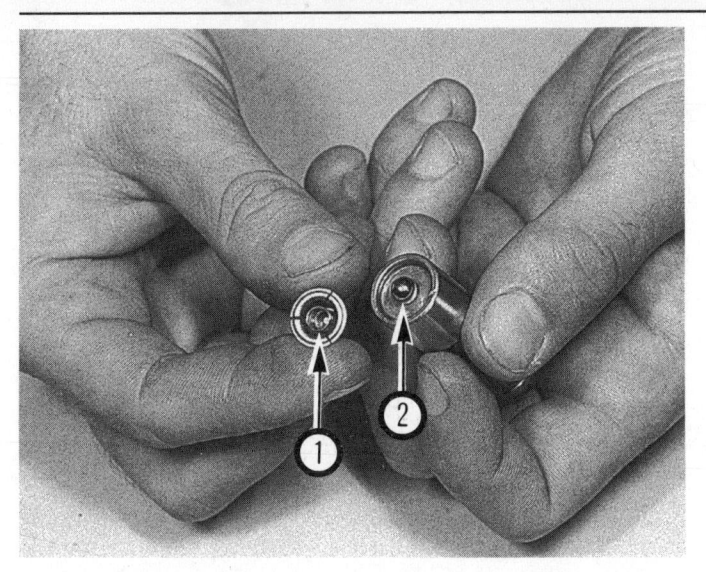

Chapter 8 Supplement: Revisions and information on later models 113

4.44A Spring (1) located in plunger cap, and ball (2) located on seat in plunger

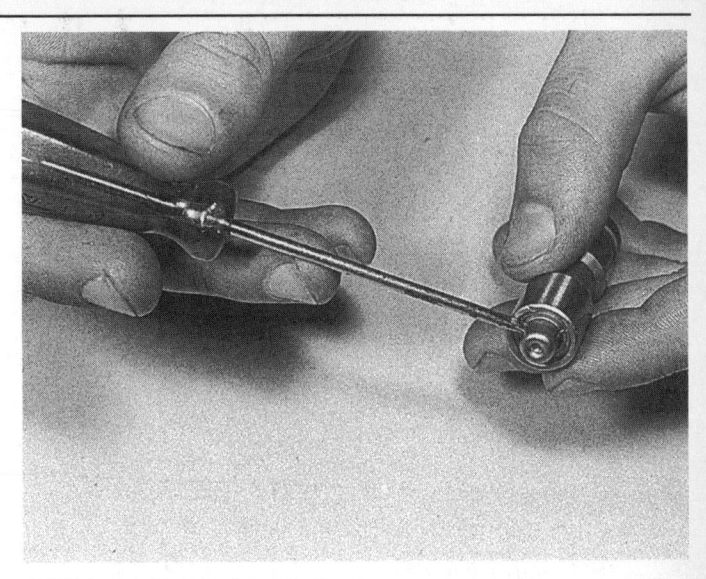

4.44B Locate the cap flange in the plunger groove

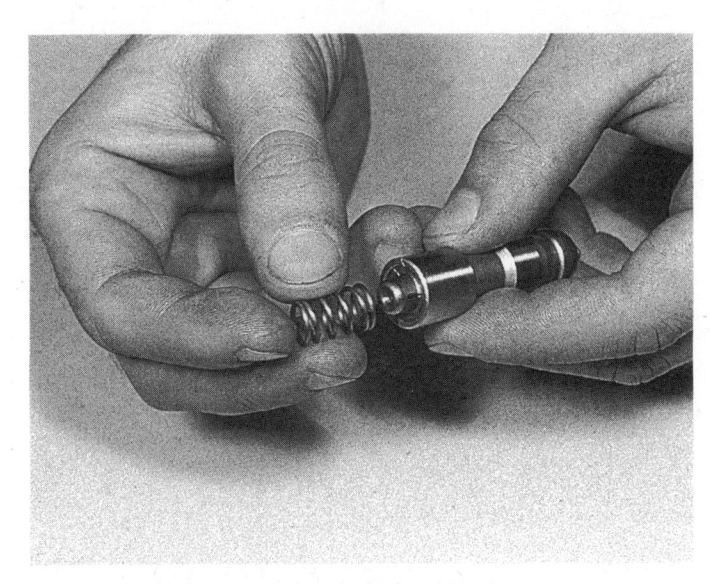

4.45A Locate the spring over the plunger cap ...

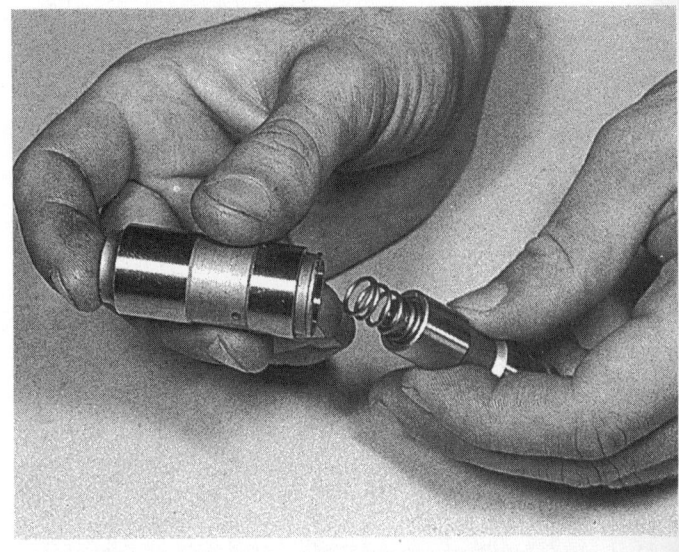

4.45B ... then slide the plunger and spring assembly into the cylinder

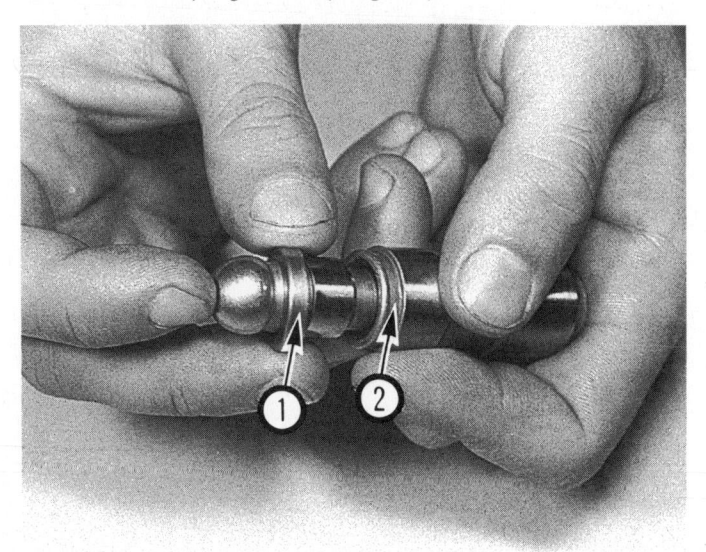

4.46 Slide the collar (1) over the top of the plunger, and engage with the groove (2) in the cylinder

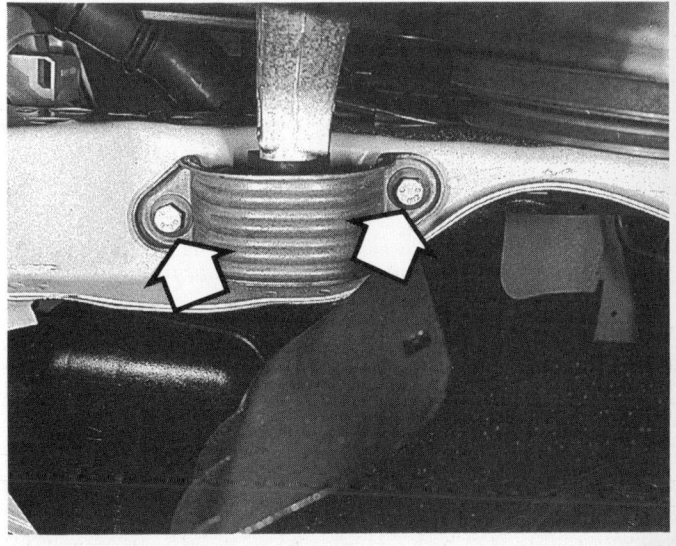

4.53 Right-hand front engine mounting-to-side chassis member securing bolts (arrowed)

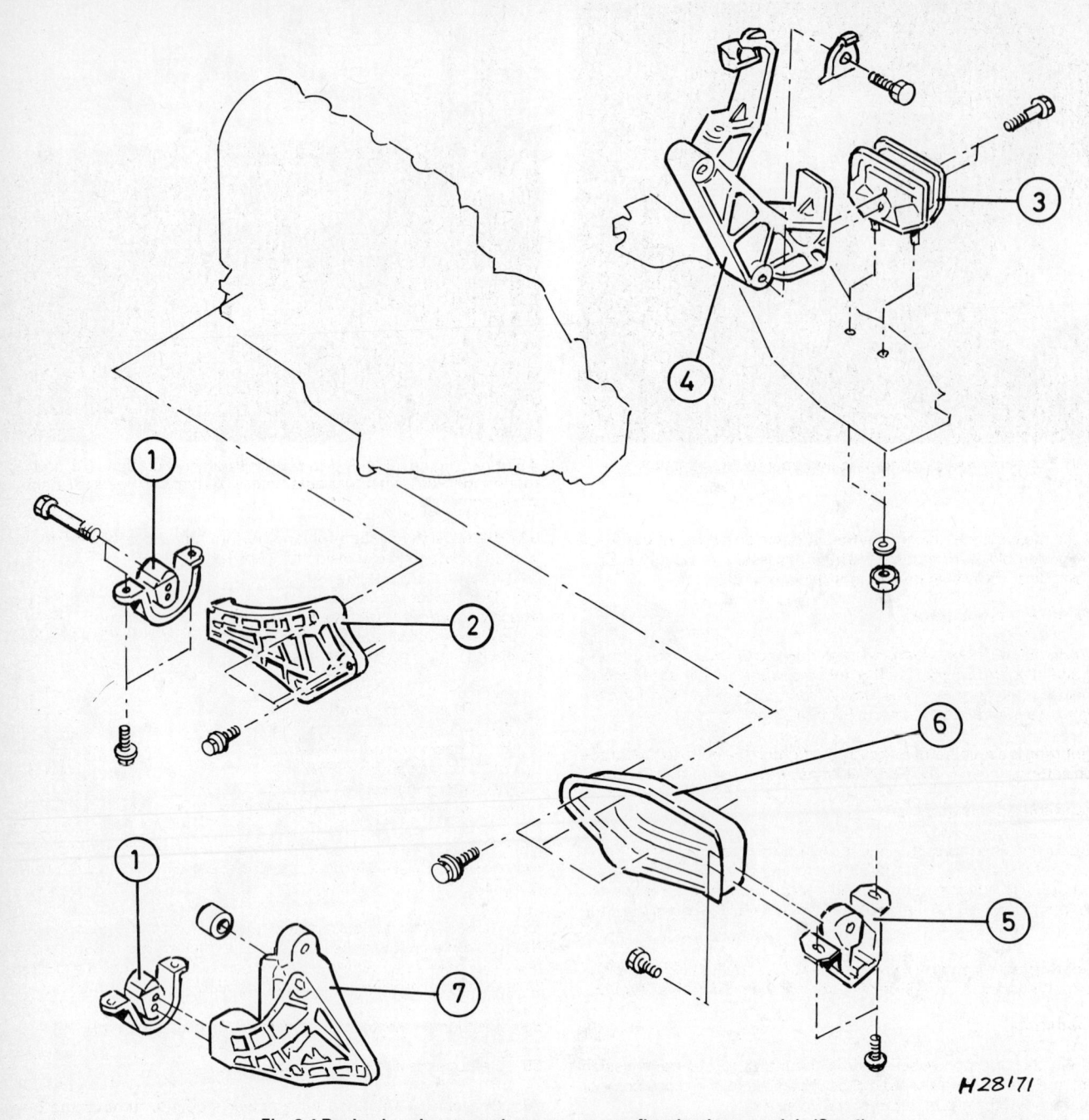

114 Chapter 8 Supplement: Revisions and information on later models

Fig. 8.4 Revised engine mounting components fitted to later models (Sec 4)

1 Right-hand front mounting
2 Right-hand front mounting bracket
3 Rear mounting
4 Rear mounting bracket
5 Left-hand front mounting
6 Left-hand front mounting bracket
7 Right-hand front mounting bracket (vehicles with power steering)

Cylinder bores – examination and renovation

48 Since this manual was originally published, the manufacturers have announced the availability of cylinder liners for certain of the ohc diesel engines covered. The fitting of these liners requires full engine reconditioning facilities, and the work must therefore be entrusted to a Vauxhall/Opel dealer or engine reconditioning specialist.

49 The introduction of liners does offer a new lease of life to an engine with worn cylinder bores which has already been rebored to the maximum permissible oversize. Note, however, that it is not permissible to fit oversize pistons to an engine which has been fitted with liners. The makers state that the wall thickness is such that boring them oversize will weaken them too much. It follows that once liners have been fitted, the only way to reclaim the block subsequently would be to fit new liners and bore these back to the standard diameter again.

Engine/transmission mountings (17D and 17DR engines) – renewal

50 A revised engine mounting arrangement is used on the 17D and 17DR engines. Removal and refitting procedures are virtually identical for all models, and are as follows.

51 It is possible to remove and refit the front mountings from above, but access is improved if the work is carried out from below. If so, raise and securely support the front of the vehicle on axle stands.

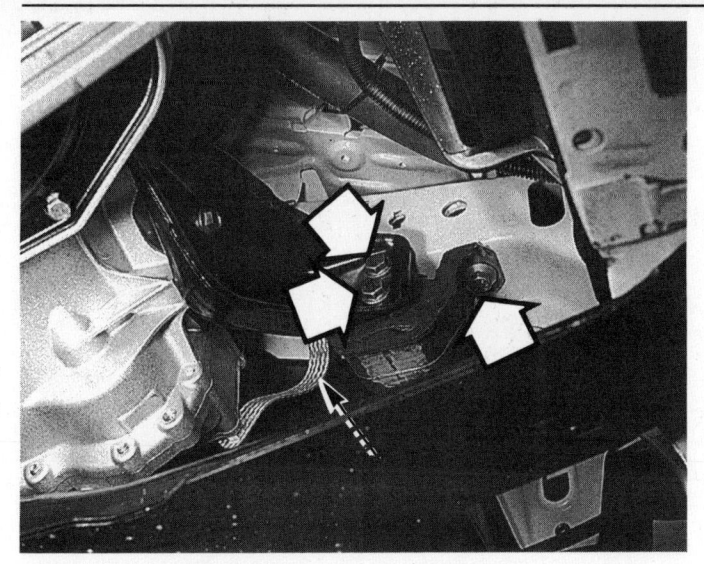

Chapter 8 Supplement: Revisions and information on later models

4.55 Left-hand front engine mounting and transmission bracket attachments (arrowed)

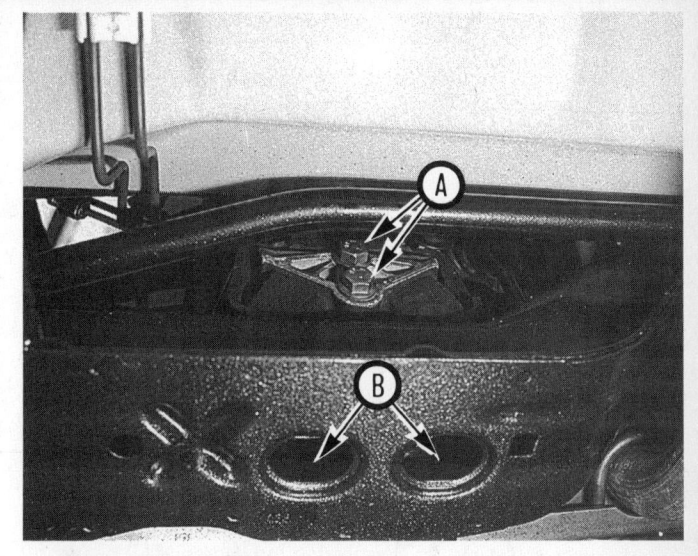

4.57 Rear engine mounting-to-transmission bracket bolts (A), and mounting-to-crossmember nuts accessible through holes (B) in the crossmember

52 Take the weight of the engine/transmission on a hoist, or use a jack and a wooden block from below. Where necessary, remove the inner wheel arch liner if working on the right-hand mounting.

Right-hand front mounting

53 Undo the three bolts securing the mounting bracket to the cylinder block, and the two bolts securing the mounting to the side chassis member (photo). Remove the assembly from the vehicle, then undo the two bolts to separate the mounting from the bracket.

54 Refitting is a reversal of removal, but tighten the bolts to the torque wrench settings given in the Specifications at the start of this Chapter.

Left-hand front mounting

55 Undo the two bolts securing the mounting to the transmission bracket, and the two bolts securing the mounting to the side chassis member (photo). Note the transmission earth strap secured by one of the mounting bolts (where applicable). Remove the mounting from the vehicle.

56 Refitting is a reversal of removal, but tighten the bolts to the torque wrench settings given in the Specifications at the start of this Chapter.

Rear mounting

57 Undo the two bolts securing the mounting to the transmission bracket, and the two nuts securing the mounting to the crossmember (photo). Remove the mounting from the vehicle.

58 Refitting is a reversal of removal, but tighten the bolts to the torque wrench settings given in the Specifications at the start of this Chapter.

Engine (17D and 17DR) – removal and refitting

59 On the 17D and 17DR units, the recommended method of engine removal is from above, leaving the transmission in the vehicle. It will be necessary to raise and support the front of the vehicle for some of the following operations. Ensure adequate means of support are used when the vehicle is raised.

60 Disconnect and remove the battery.

61 Remove the bonnet.

62 Remove the air cleaner housing and the air intake hoses.

63 Drain the cooling system, taking precautions against scalding if the coolant is hot. See Chapter 2, Section 3.

64 Remove the radiator top and bottom hoses completely, together with the heater and expansion tank hoses.

65 Disconnect all the electrical leads to the engine, making notes or attaching labels if there is any possibility of confusion later. The wiring to be disconnected is largely dependent on model, but will include the following:

(a) Oil pressure switch.
(b) Alternator.
(c) Coolant temperature sender.
(d) Injection pump wiring.
(e) Glow plug feed to bus bar.
(f) Radiator fan and thermoswitch.

66 Removal of the radiator and cooling fan is recommended, both to improve access and to reduce the risk of damage. Refer to Chapter 2, Section 6.

67 Clean around the fuel inlet and return unions at the injection pump, then disconnect them. Be prepared for fuel spillage. Cap the open unions (eg with polythene and rubber bands) to keep fuel in and dirt out. Detach the fuel hoses from their engine clips.

68 Disconnect the throttle cable from the injection pump.

69 Remove the alternator drivebelt.

70 On vehicles equipped with power steering, remove the power steering pump from the engine, and place it to one side without disconnecting any of the fluid hoses (refer to the relevant petrol-engine manual).

71 Disconnect the brake servo vacuum pump hose at the vacuum pump.

72 Undo the four bolts and remove the crankshaft pulley.

73 Remove the starter motor.

74 Remove the clutch housing cover plate from the bottom of the bellhousing.

75 On automatic transmission models, unbolt the torque converter from the driveplate. Turn the crankshaft to gain access to the bolts through the aperture in the bellhousing.

76 Unbolt the exhaust downpipe from the manifold flange and the flexible joint, and remove it from the vehicle.

116   Chapter 8 Supplement: Revisions and information on later models

77   Remove the oil filter.

78   Attach suitable lifting gear to the engine, and just take the weight of the unit.

79   Undo all the bolts securing the engine to the transmission bell-housing.

80   Remove the right-hand engine mounting, together with the mounting bracket, from the cylinder block.

81   Make a final check that all connections to the engine have been removed, and that there is nothing in the way that is likely to interfere when the engine is lifted out.

82   Support the transmission with a jack, and separate the engine from the transmission. Carefully lift the engine upwards and out of the engine bay, turning it as necessary to clear any obstructions. Lower the unit to the ground, and take it to the bench.

83   Refitting and reconnection of all components is a reversal of the removal sequence. Check the adjustment of all disconnected belts and cables where applicable, as described in the appropriate Chapters of this manual (or the relevant petrol-engine manual). Tighten all bolts to the correct torque wrench settings (where given) and refill the cooling system on completion.

6.5 Heated fuel filter showing fuel temperature sensor (A) and heater unit (B)

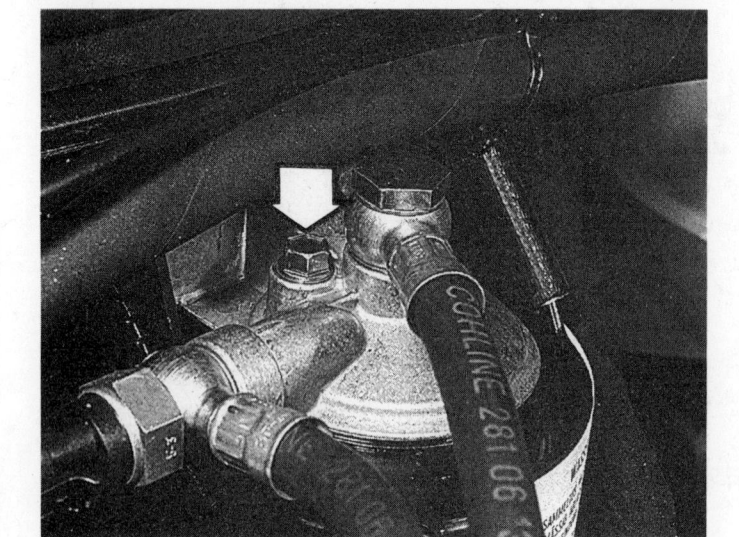

5 Cooling system

Water pump (17D and 17DR engines) – removal and refitting

1   Water pump removal and refitting procedures on the 17D and 17DR engines are essentially the same as described in Chapter 2, but due to the modifications to the camshaft drivebelt covers and tensioner, bear in mind the following.

2   Remove the camshaft drivebelt as described in Section 4 of this Chapter.

3   On the 17D engine, remove the additional bolt below the water pump which secures the camshaft drivebelt inner cover. On the 17DR engine, remove the camshaft drivebelt automatic tensioner after the drivebelt has been removed.

4   On the 17DR engine, when refitting the water pump and prior to refitting the drivebelt, first ensure that the pump is correctly positioned by checking that the lug on the water pump flange is aligned with the corresponding lug on the cylinder block.

5   Refit and tension the camshaft drivebelt according to the procedures given in Section 4 of this Chapter.

6.6A Top of revised fuel filter unit, showing vent plug location (arrowed)

6 Fuel and exhaust systems

Maintenance and inspection

1   The fuel system remains largely unchanged from that described in Chapter 3. However, there are detail enhancements and modifications to models fitted with the 17D and 17DR engines, these being described in detail below.

2   There have been a number of cases where fuel system faults have been incorrectly attributed to injection pump failures, when the real problem was due to leaks in the system in general. In most instances, this has become apparent on models fitted with Lucas/CAV injection pumps, but the general checks described here can be applied equally to Bosch-equipped models.

6.6B Fuel filter assembly removed from the bulkhead to show position of drain plug (A) and hose (B)   arrowed

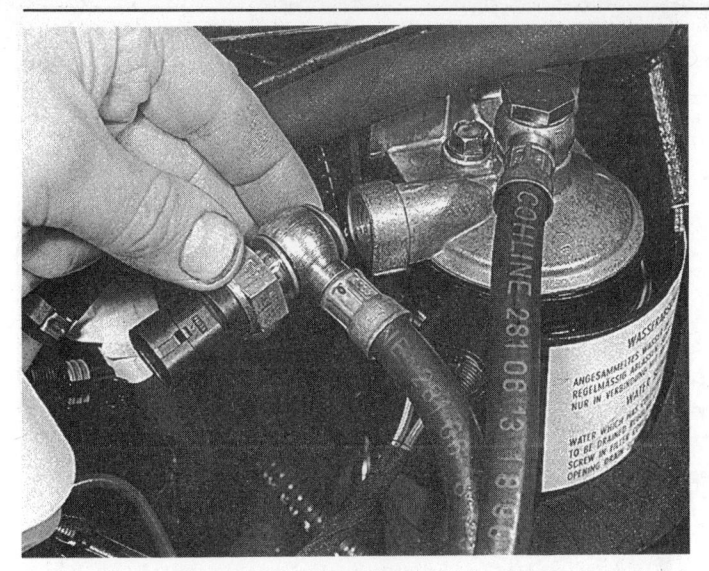

Chapter 8 Supplement: Revisions and information on later models

6.7A Fuel temperature sensor is combined with hose union banjo bolt

6.7B Filter heater is switched by this relay. Note 30A fuse in the background (arrowed)

6.7C Electrical ancillary box also houses glow plug relay and its 60A fuse

6.8 Revised air cleaner filter element housed in rectangular housing

3   In cases where general rough running or engine misfiring is occurring, check that loose or damaged fuel lines or connections are not allowing air to enter the fuel system. This is most easily done with the vehicle raised on a ramp or axle stands to permit visual inspection of the entire fuel system. Apply low air pressure to the fuel tank. The manufacturer offers no suggestions as to how this might be achieved, but one method would be to seal the fuel tank filler, then connect a footpump to one of the breather pipe stubs on the fuel tank. Apply gentle pressure to the system, and check carefully for signs of leakage.

4   Any leakage of fuel discovered should be investigated and the fault rectified. Since the fuel flow line is under slight negative pressure in normal use, fuel leakage while the system is under pressure is likely to equate with air bubbles being drawn into the fuel being delivered to the pump during normal operation. It is also worth remembering that a loose connection on the line from the tank to the pump could also allow surface water to be drawn into the system – check for this if the filter has shown signs of unusually high water contamination.

Heated fuel filter
5   On some later 16D engines, and all 17D and 17DR versions, an electric heating element is fitted between the filter housing and bowl. A sensor monitors the fuel temperature passing through the filter, and if this falls to a point where fuel waxing is likely to occur, the heater is switched on to warm the fuel (photo).

6   For most purposes, the filter can be dealt with as described in Chapter 3, Sections 2 and 3. It is easier to release the assembly from the bulkhead to carry out the periodic draining operation, as access below the filter is restricted while it is in place, making the operation potentially messy. The main reason for the poor access problem is the adoption of the heat shield around the filter head (photos).

7   The temperature sensor is combined in a modified banjo bolt which passes through one of the fuel line unions and secures it to the filter head. The heater is switched by a relay, which is housed together with its 30A supply fuse in the electrical ancillary box on the left-hand side of the bulkhead (photos). No details of test procedures or specifications are supplied by the manufacturers, but in the event of a suspected fault, check the sensor, relay and the heater unit by substitution of new units.

Air cleaner – element renewal
8   As mentioned earlier in this Chapter, a revised intake system designed to reduce noise is fitted to later models. The air cleaner element has been relocated to a rectangular plastic air cleaner housing, mounted on the inner wing just behind the right-hand headlight. Access to the element is gained after the clips securing the air cleaner lid have been released (photo).

118 Chapter 8 Supplement: Revisions and information on later models

Fig. 8.5 Exploded view of typical later-type remote air cleaner assembly - Cavalier model shown (Sec 6)

1 Air cleaner lower casing
2 Filter element
3 Air cleaner/resonator tube
4 Resonator
5 Sealing ring
6 Screw
7 Intake gasket
8 Rubber mounting
9 Air inlet hose
10 Hose clips

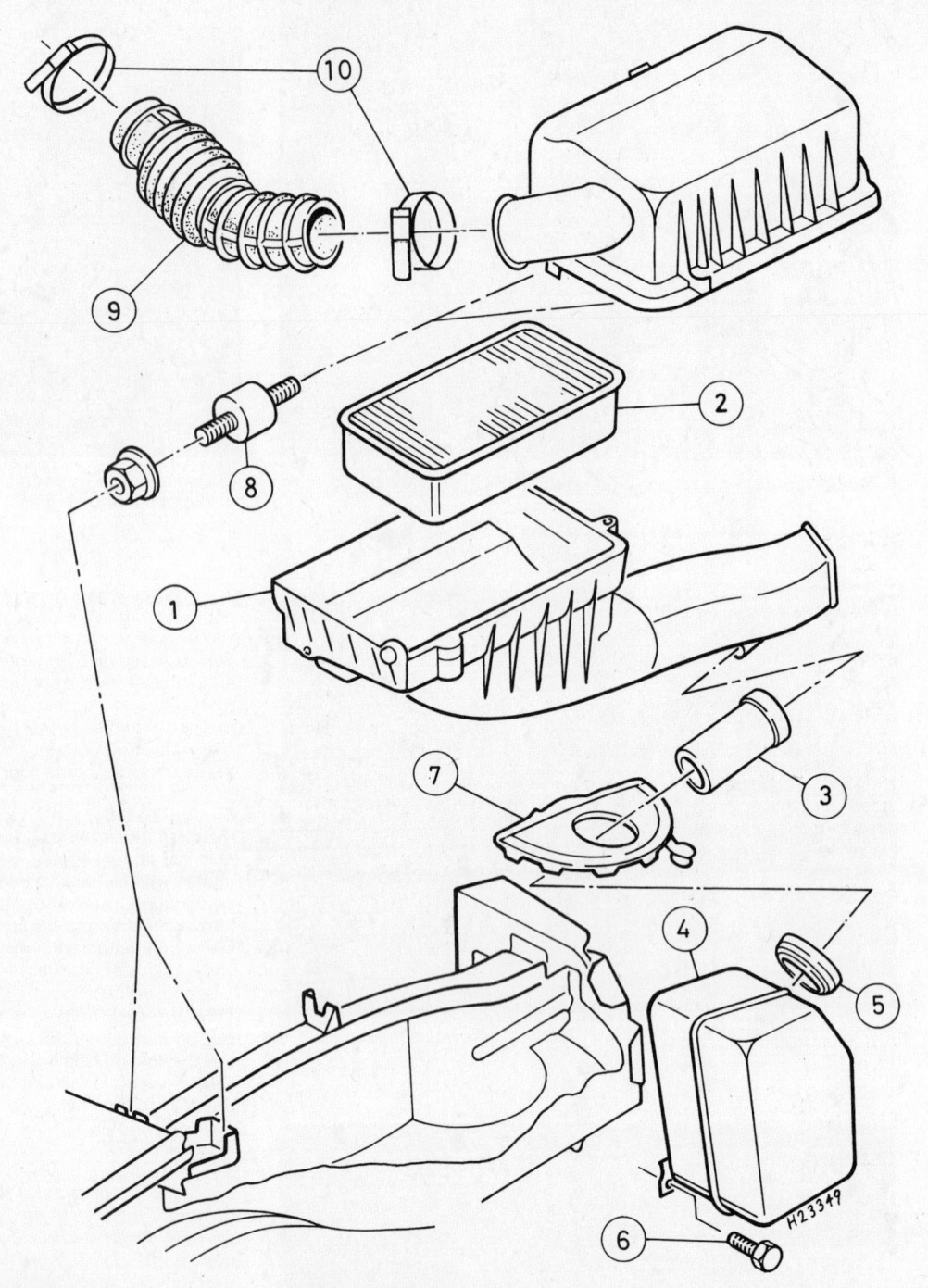

Air cleaner housing – removal and refitting

9   To remove the later-type remote air cleaner housing, first unclip and withdraw the air cleaner lid, and lift out the element.

10   Where fitted, slacken the hose clip, and release the air intake hose from the front of the housing.

11   Undo the screws or slacken the two nuts, according to model, and lift away the housing.

12   On certain models, it may be necessary to remove the headlight unit to allow sufficient clearance to disengage the filter snorkel that runs beneath the headlight unit. Refer to the relevant petrol-engine manual for details.

13   Refitting is a reversal of removal.

Idle and maximum speed adjustment – Bosch VE fuel injection pump

Later 16DA and all 17D engines

14   Later versions of the Bosch VE injection pump require a revised idle speed adjustment procedure to that described in Chapter 3, Section 5. The later pump can be identified by the vertical idle speed adjustment screw located on the front face of the pump. Where this type of pump is fitted, it is essential that the engine speed control lever stop screw is not disturbed. The positions of the adjustment screws are shown in Fig. 8.6.

Caution: *The manufacturer warns that pumps which have had the stop screw tampered with cannot be re-adjusted by a dealer, which suggests that the pump would have to be set up by a Bosch specialist.*

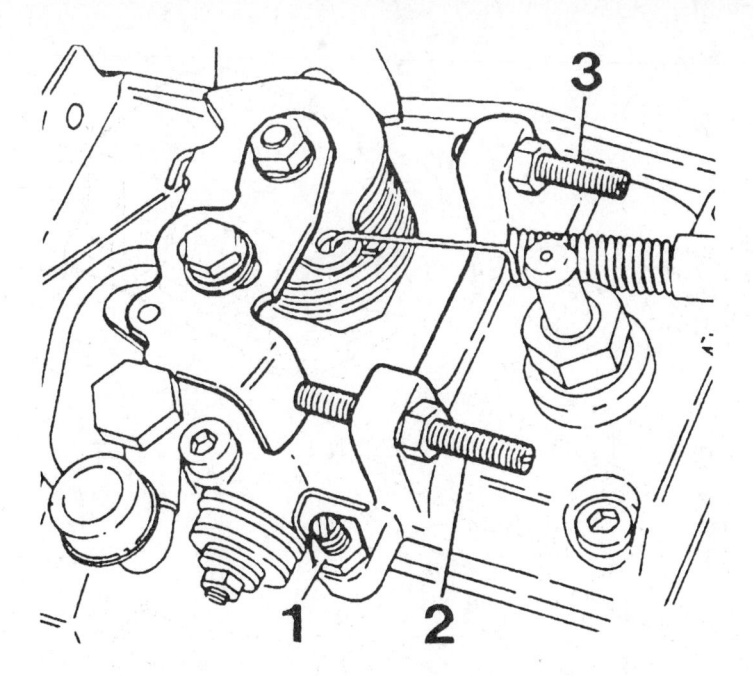

Chapter 8 Supplement: Revisions and information on later models

Fig. 8.6 Idle and maximum speed adjustment points – Bosch VE injection pump, 16DA and 17D engines (Sec 6)

1 Idle speed adjustment screw
2 Engine speed control lever stop screw – do not disturb
3 Maximum speed adjustment screw

15 Refer to Chapter 3, Section 5, for details of various methods of measuring the engine speed. It is recommended that the methods described in paragraph 1 or 3 of that Section be adopted for the following adjustments.

16 To adjust the idle speed, slacken the locknut on the idle speed adjustment screw, and turn the screw as necessary to obtain the specified speed. Tighten the locknut, without disturbing the screw position, on completion.

17 The procedure for adjustment of the maximum speed is the same as described in Chapter 3, Section 6. Note, however, that these later injection pumps have their maximum speed adjustment screws locked with a lead seal which must be removed for adjustment. As the screw should ideally be resealed after adjustment, it may be beneficial to leave this operation to a Bosch injection specialist.

17DR engines

18 The Bosch injection pump fitted to the 17DR engine is equipped with a vacuum-operated cold start device which allows the engine idle speed to increase when the temperature is below 20°C (68°F) (photo). Two idle speed adjustments are therefore necessary – one for a cold engine and one for a hot engine.

19 Refer to Chapter 3, Section 5, for details of various methods of measuring the engine speed. It is recommended that the methods described in paragraph 1 or 3 of that Section be adopted for the following adjustments.

20 With the engine cold, ie below 20°C (68°F), check that there is approximately 2.0 to 3.0 mm (0.08 to 0.12 in) of free play between the clamping sleeve on the end of the cold start device operating cable, and the actuating lever (Fig. 8.7). Alter the position of the clamping sleeve if necessary.

21 Start the engine, and check that the cold idling speed is as given in the Specifications at the start of this Chapter. If adjustment is necessary, turn the cold idling speed adjustment screw (Fig. 8.8) as required to achieve the desired speed.

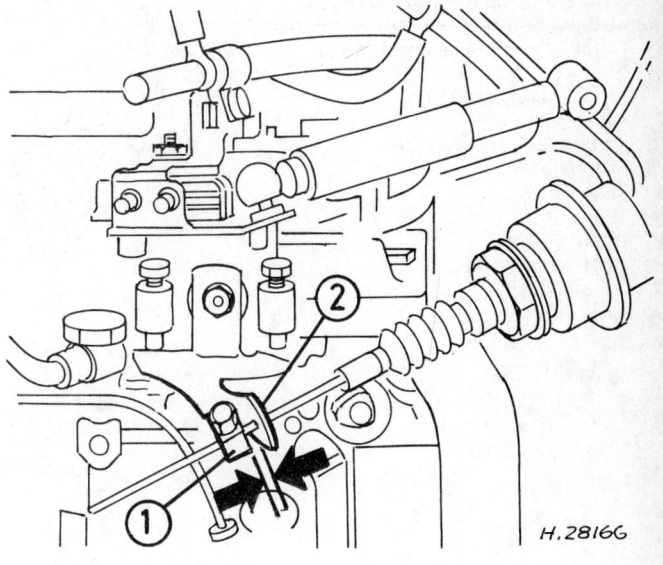

Fig. 8.7 Checking cold start device free play – Bosch VE injection pump, 17DR engines (Sec 6)

1 Clamping sleeve
2 Actuating lever

Arrows indicate free play checking point – engine cold

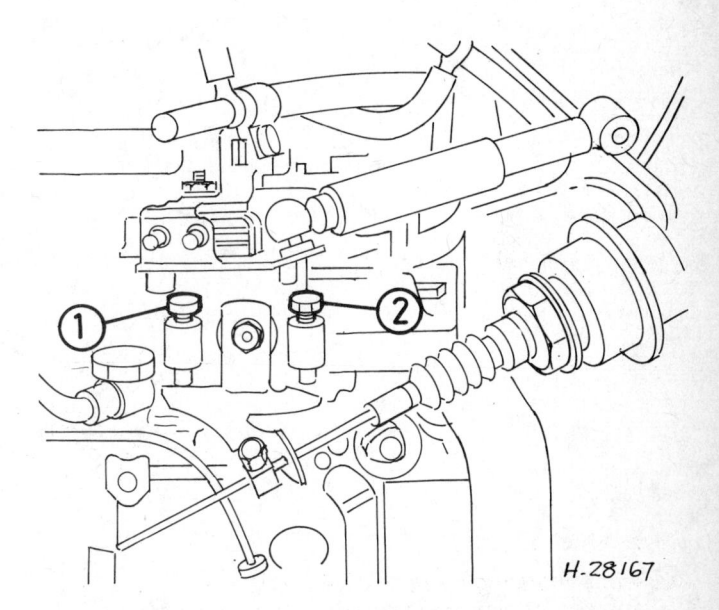

Fig. 8.8 Idle speed adjustment screws – Bosch VE injection pump, 17DR engines (Sec 6)

1 Cold idling speed adjustment screw
2 Hot idling speed adjustment screw

22 Warm the engine up by taking the vehicle on a short test drive. With the engine temperature above 20°C (68°F), check that the cold start device operating cable has retracted and moved the actuating lever into contact with the hot idling speed adjustment screw (Fig. 8.8). The engine should now be idling at the (lower) hot engine idling speed as given in the Specifications. If adjustment is necessary, turn the hot idling adjustment screw as necessary.

Chapter 8 Supplement: Revisions and information on later models

6.18 Vacuum-operated cold start device (arrowed) fitted to the later Bosch injection pump

23 The procedure for adjustment of the maximum speed is the same as described in Chapter 3, Section 6. Refer to Fig. 8.9 for the location of the adjustment screw. However, these later injection pumps may have their maximum speed adjustment screws locked with a lead seal which must be removed for adjustment. As the screw should ideally be resealed after adjustment, it may be beneficial to leave this operation to a Bosch injection specialist.

Idle and maximum speed adjustment – Lucas/CAV fuel injection pump

Later 16DA and all 17D engines

24 The idle speed on engines fitted with the Lucas/CAV injection pump is adjusted using the stop screw shown in Fig. 8.10. The engine speed control lever stop screw must not be disturbed, and it is covered with a plastic anti-tamper cap to indicate this.

25 Refer to Chapter 3, Section 5, for details of various methods of measuring the engine speed. It is recommended that the methods described in paragraph 1 or 3 of that Section be adopted for the following adjustments.

26 To adjust the idle speed, slacken the locknut on the idle speed stop screw, and turn the screw as necessary to obtain the specified speed. Tighten the locknut, without disturbing the screw position, on completion.

27 Maximum (cut-off) speed is set in production, using the cut-off speed stop screw (Fig. 8.10). The screw is sealed with lead after adjustment has been made. As with all injection pumps it is not normally necessary to disturb the cut-off speed setting in normal circumstances, but if adjustment is necessary, it is recommended that this be carried out by a Lucas/CAV injection specialist.

17DR engines

28 As with the Bosch pump described earlier, the Lucas/CAV injection pump fitted to the 17DR engine is also equipped with a vacuum-operated cold start device which allows the engine idle speed to increase when the temperature is below 20°C (68°F). Two idle speed adjustments are therefore necessary — one for a cold engine and one for a hot engine.

29 Refer to Chapter 3, Section 5, for details of various methods of measuring the engine speed. It is recommended that the methods described in paragraph 1 or 3 of that Section be adopted for the following adjustments.

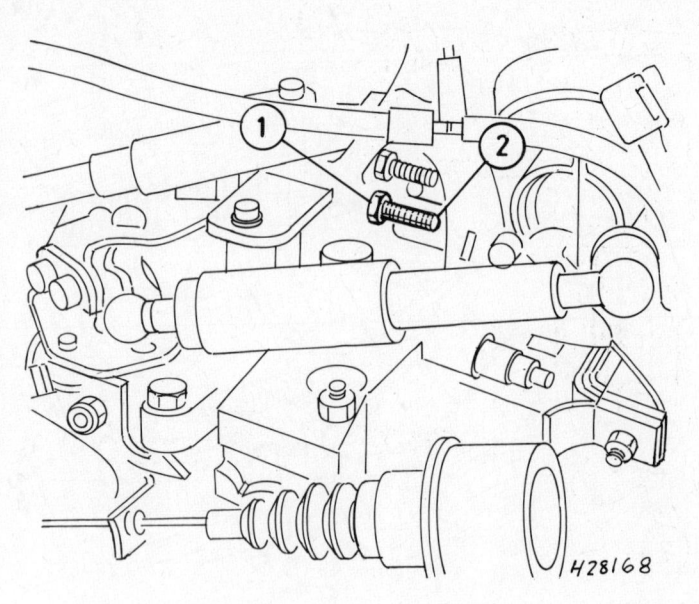

Fig. 8.9 Maximum speed adjustment screw location – Bosch VE injection pump, 17DR engines (Sec 6)

1 Locknut   2 Adjustment screw

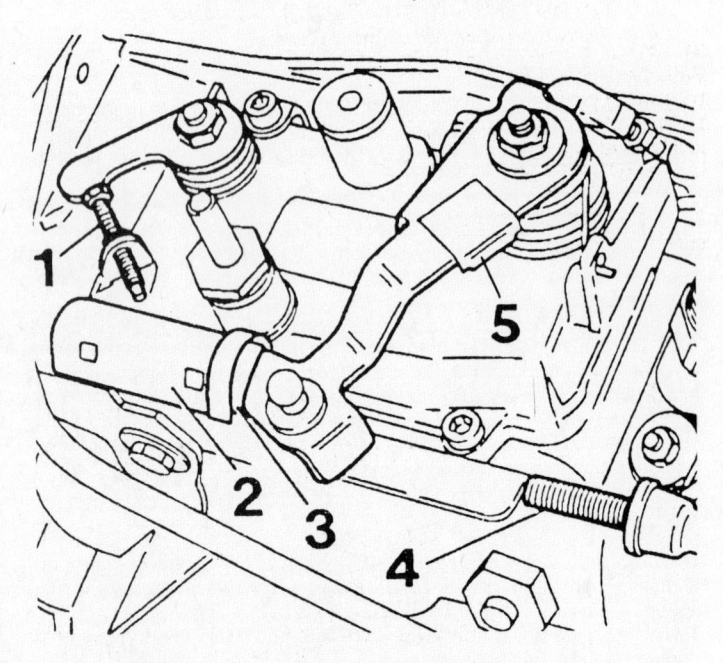

Fig. 8.10 Idle and maximum speed adjustment points – Lucas/CAV injection pump, 16DA and 17D engines (Sec 6)

1 Idle speed stop screw
2 Plastic anti-tamper cap
3 Engine speed control lever stop screw – do not disturb
4 Cut-off speed stop screw
5 Timing value for individual pump (marked on plate)

30 With the engine cold, ie below 20°C (68°F), check that the cold idling speed is as given in the Specifications at the start of this Chapter. If adjustment is necessary, slacken the locknut on the vacuum unit thrust-rod, and turn the thrust-rod as required to achieve the desired speed (Fig. 8.11). Tighten the locknut when the setting is correct.

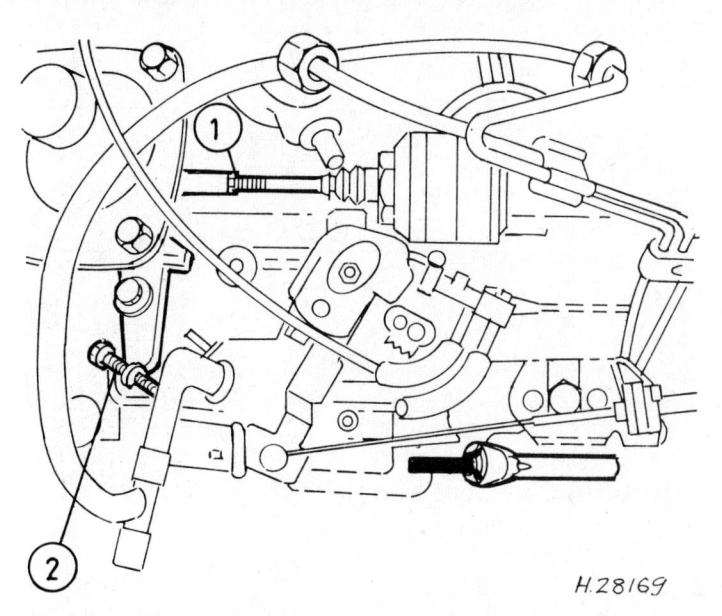

Chapter 8 Supplement: Revisions and information on later models

Fig. 8.11 Idle speed adjustment screws – Lucas/CAV injection pump, 17DR engines (Sec 6)

1 Vacuum unit thrust-rod locknut (cold idle adjustment)
2 Hot idling speed adjustment screw

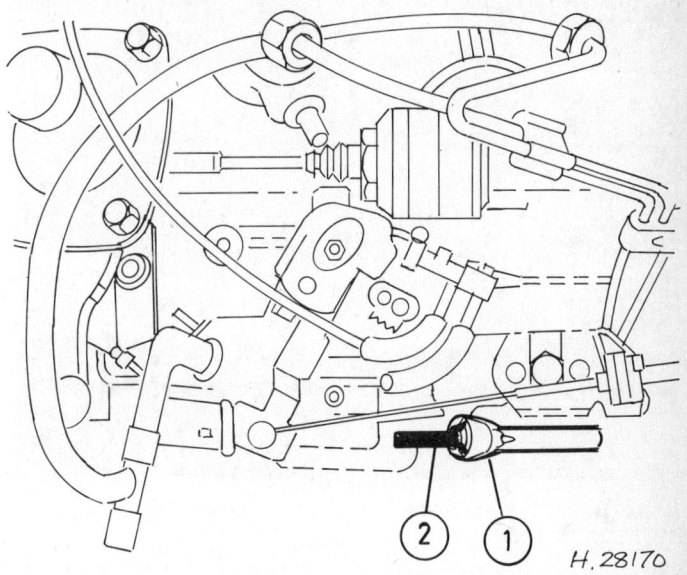

Fig. 8.12 Maximum speed adjustment screw location – Lucas/CAV injection pump, 17DR engines (Sec 6)

1 Lead seal   2 Cut-off speed stop screw

31 Warm the engine up by taking the vehicle on a short test drive. With the engine temperature above 20°C (68°F), the engine should now be idling at the (lower) hot engine idling speed as given in the Specifications. If adjustment is necessary, slacken the locknut and turn the hot idling speed adjustment screw as necessary (Fig. 8.11). Tighten the locknut when the setting is correct.

32 Maximum (cut-off) speed is set in production, using the cut-off speed stop screw (Fig. 8.12). The screw is sealed with lead after adjustment has been made. As with all injection pumps, it is not normally necessary to disturb the cut-off speed setting in normal circumstances; if adjustment is thought necessary, it is recommended that this be carried out by a Lucas/CAV injection specialist.

Fuel injection pump timing (17D and 17DR engines) – checking and adjustment

33 On later engines, the pump timing procedures are generally similar to those given in Chapter 3, Section 7, but noting the following points.

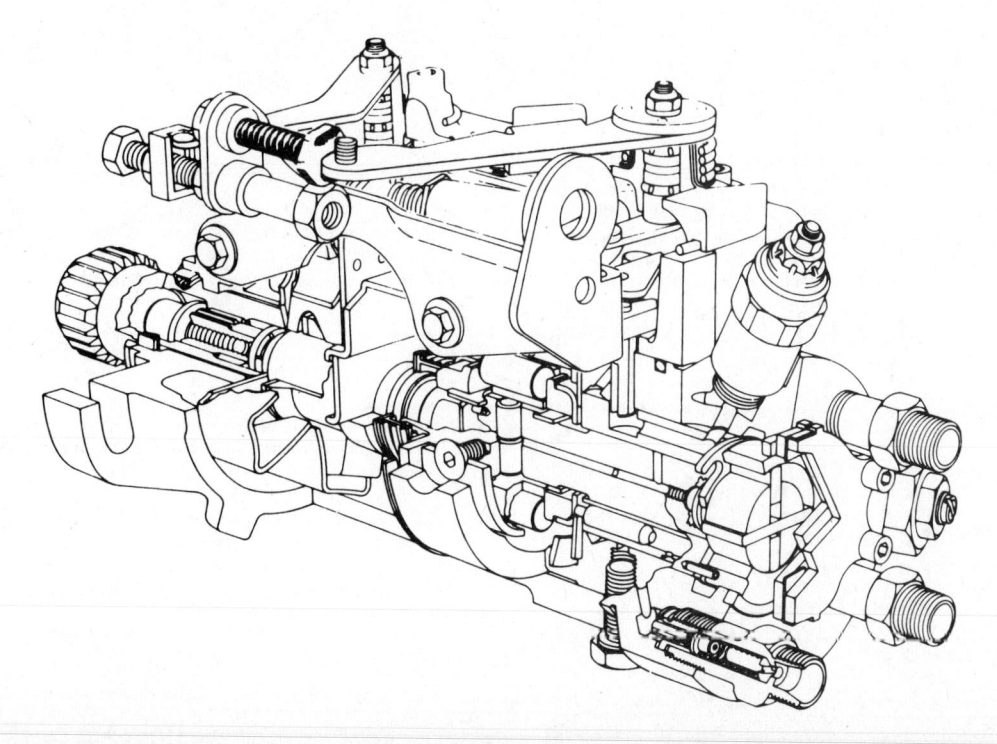

Fig. 8.13 Part-sectional view of Lucas/CAV fuel injection pump fitted to 17D engines (Sec 6)

Chapter 8 Supplement: Revisions and information on later models

6.34A Flywheel timing marks are visible through clutch housing inspection cover

6.34B Pump pulley timing mark aligned with moulded mark on the drivebelt inner cover

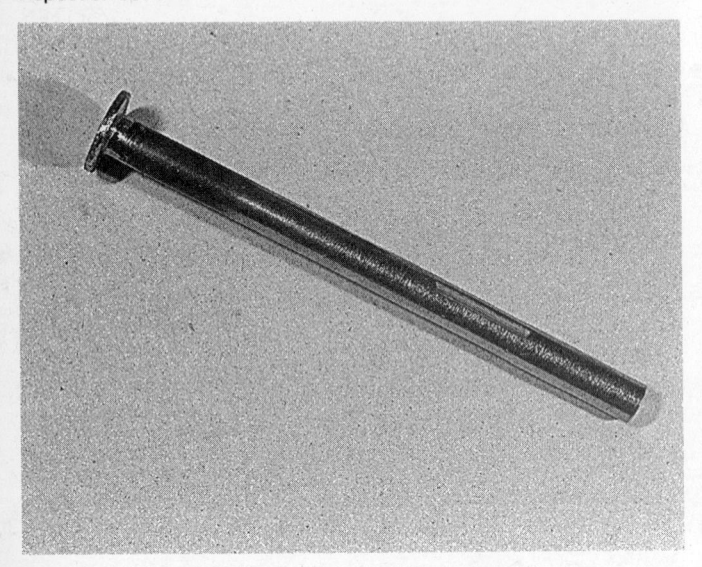

6.36 Home-made probe used for checking Lucas/CAV pump timing

6.37 Lucas/CAV pump showing DTI set up for timing check. Individual value for each pump is stamped on plate (arrowed)

34 With the introduction of the revised camshaft drivebelt inner and outer covers, the timing mark for the injection pump sprocket is now located on the drivebelt inner cover. With No 1 piston set to TDC on the firing stroke, the TDC mark on the flywheel and the pointer on the clutch housing will be aligned, and the timing mark on the injection pump pulley will be aligned with the moulded mark on the drivebelt inner cover (photos).

35 There are also some slight changes to the timing procedure when dealing with the Lucas/CAV injection pump as detailed below.

36 When working on the Lucas/CAV injection pump, note that the closing plug is located on the upper surface of the pump rather than at the end of the pump casing as it is on the Bosch pump. In the absence of the measuring tool KM-690-A and the dial test indicator KM-571-B, you will need a standard dial test indicator (DTI), together with some method of mounting it above the timing hole at the appropriate height. Also required is a headed probe made to the dimensions shown in Fig. 8.14, this being placed in the timing hole before the DTI is mounted in position (photo).

37 Check the amount of lift indicated on the DTI when the crankshaft timing marks are brought into alignment. There is no standard specified lift figure for Lucas/CAV pumps - each pump is calibrated during manufacture, and the lift figure marked on a plate which is fitted to the pump lever (photo). If the lift figure shown on the DTI does not correspond with that given on the plate, adjust the pump sprocket as described in Chapter 3, Section 7. Once adjustment is complete, remove the DTI and probe, and refit the closing plug.

Vacuum-operated cold start device (17DR engines) – removal and refitting

Bosch fuel injection pump

38 Slacken the lockbolt, and remove the clamping sleeve from the end of the cold start device operating cable.

39 Disconnect the vacuum hose, undo the clamping nut, and remove the cold start device from its mounting bracket.

40 Refitting is a reversal of removal. On completion, adjust the idle speed as described earlier in this Section.

Lucas/CAV injection pump

41 Undo the two bolts securing the cold start device mounting bracket to the injection pump. Note the position of the end of the speed control lever return spring.

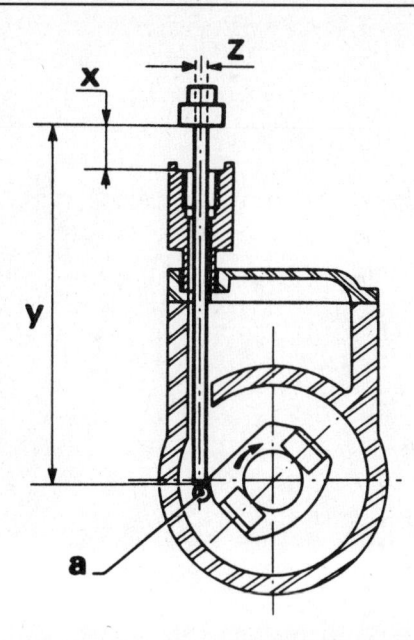

Chapter 8 Supplement: Revisions and information on later models

Fig. 8.14 Special DTI probe shown in position during pump timing check – Lucas/CAV injection pump (Sec 6)

- a Timing piece
- x Timing value (as shown on plate)
- y 95.5 ± 0.01 mm
- z 7.00 mm shank diameter

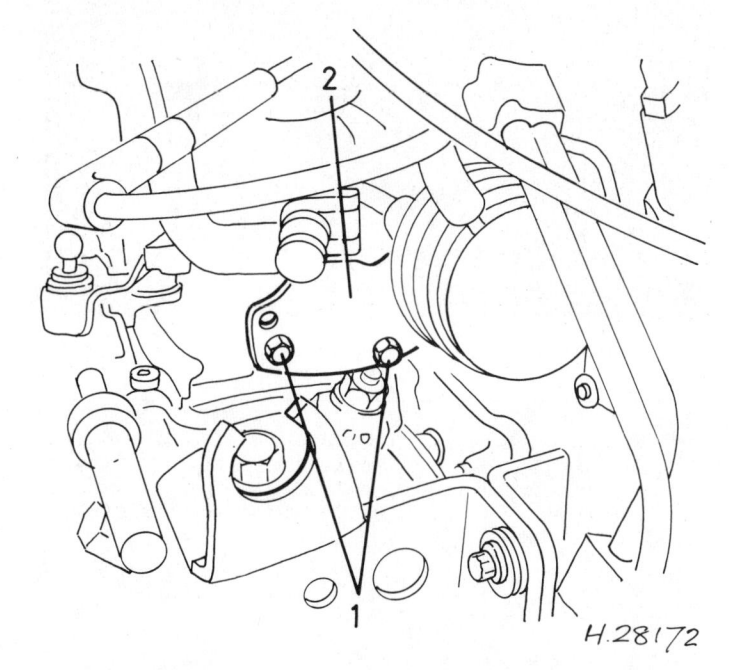

Fig. 8.15 Lucas/CAV injection pump cold start device attachments – 17DR engines (Sec 6)

1. Mounting bracket securing bolts
2. Mounting bracket

42 Release the mounting bracket, then detach the vacuum hoses from the cold start device, noting their locations.

43 Disconnect the cold start device thrust rod, and remove the unit complete with mounting bracket. Separate the device from the mounting bracket if necessary, after removal.

6.46 Using a glow plug tester to check current draw and heating time of a glow plug

44 Refitting is a reversal of removal. On completion, adjust the idle speed as described earlier in this Section.

Preheating system – notes on testing

Note: *The following information is applicable only to the 11-volt glow plugs fitted to 16D, 16DA and early 17D engines. Later engines have a 5-volt preheating system, which cannot be tested without specialist equipment. Ensure that 11-volt glow plugs are fitted to your engine before carrying out the following tests.*

Caution: *Glow plug tips become very hot during testing. Take care to avoid burns, especially from plugs which have only just stopped visibly glowing.*

45 In addition to the checks and tests described in Chapter 3, it can be useful to carry out a visual operational check of a suspect glow plug.

46 There are commercially-produced glow plug testers available, which comprise a casing in which the plug is clamped, an ammeter, 12-volt connection leads, and a simple timing circuit which illuminates successive LEDs in five-second intervals. With the glow plug in place and the leads connected to a 12-volt battery, note the time taken before the tip of the plug begins to glow and the current drops (photo).

47 Whilst such equipment is available in a commercial workshop, it is too infrequently needed by most owners to justify the purchase price. However, an equivalent test rig can be made up at home at little cost.

48 Using an ammeter with a range of at least 30A, connect a lead to each terminal, fitting a crocodile clip at each end. In the interests of safety, fit an in-line fuseholder to one of the leads, using a 30A fuse.

49 With the suspect glow plug removed from the cylinder head, clamp it with self-locking pliers against its metal body to a sound earth point on the engine. Connect the lead from the positive (+) terminal on the ammeter to the glow plug terminal (photo). Have an assistant ready with a watch so that a running count of seconds elapsed can be made while you watch the glow plug tip.

50 Connect the remaining crocodile clip from the ammeter negative (–) terminal to the battery positive terminal, and start the count. Watch the ammeter needle and the glow plug tip closely. The glow plug tip should start to glow red after about five seconds; after about fifteen seconds, the current reading should drop from around 25A to about 12A.

51 Note that the above timings and current figures are not precise. If the glow plug under test performs reasonably closely to the above sequence, it is likely to be in serviceable condition. An abnormally high or low current reading (or a blown fuse) indicates the need for renewal, as does a failure of the tip to glow at all.

6.49 Home-made glow plug test rig in use. Note glow plug clamped to a sound earth point (arrowed)

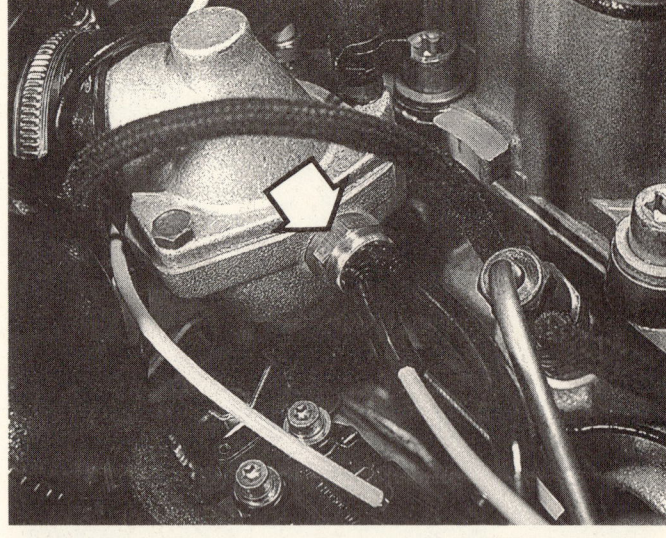

6.53A Thermo-switch (arrowed) for control of the EGR system vacuum supply

6.53B Additional EGR system vacuum take-off (arrowed) on vacuum pump

6.54A EGR valve (arrowed) mounted on inlet manifold

6.54B EGR system corrugated pipe (arrowed) between exhaust and inlet manifolds

Exhaust gas recirculation (EGR) system – 17DR engines

52 The vacuum-operated EGR system reintroduces small amounts of exhaust gas into the combustion cycle, to reduce the generation of harmful exhaust pollutants.

53 The system is operational at temperatures above approximately 20°C (68°F), and when the engine is in the idle or part-load mode. A thermo-switch screwed into the thermostat housing operating, in conjunction with a vacuum switch on the injection pump, controls the supply of vacuum to the system components. The vacuum supply is provided by an additional take-off from the brake servo vacuum pump (photos). In addition, the thermo-switch also controls the vacuum supplied to the cold start device described earlier in this Section.

54 The volume of exhaust gas reintroduced is controlled by the EGR valve mounted on the inlet manifold. When vacuum is supplied to the valve, the diaphragm inside is deflected, and a bypass to the exhaust manifold is opened. Exhaust gases are then allowed into the inlet manifold, via a corrugated pipe connected between the exhaust and inlet manifolds (photos).

55 The system is virtually maintenance-free, the only routine operations necessary are checks for condition and security of the vacuum hoses and connections.

Chapter 8 Supplement: Revisions and information on later models

Conversion factors overleaf

Conversion factors

Length (distance)
Inches (in)	X	25.4	= Millimetres (mm)	X	0.0394	= Inches (in)
Feet (ft)	X	0.305	= Metres (m)	X	3.281	= Feet (ft)
Miles	X	1.609	= Kilometres (km)	X	0.621	= Miles

Volume (capacity)
Cubic inches (cu in; in^3)	X	16.387	= Cubic centimetres (cc; cm^3)	X	0.061	= Cubic inches (cu in; in^3)
Imperial pints (Imp pt)	X	0.568	= Litres (l)	X	1.76	= Imperial pints (Imp pt)
Imperial quarts (Imp qt)	X	1.137	= Litres (l)	X	0.88	= Imperial quarts (Imp qt)
Imperial quarts (Imp qt)	X	1.201	= US quarts (US qt)	X	0.833	= Imperial quarts (Imp qt)
US quarts (US qt)	X	0.946	= Litres (l)	X	1.057	= US quarts (US qt)
Imperial gallons (Imp gal)	X	4.546	= Litres (l)	X	0.22	= Imperial gallons (Imp gal)
Imperial gallons (Imp gal)	X	1.201	= US gallons (US gal)	X	0.833	= Imperial gallons (Imp gal)
US gallons (US gal)	X	3.785	= Litres (l)	X	0.264	= US gallons (US gal)

Mass (weight)
Ounces (oz)	X	28.35	= Grams (g)	X	0.035	= Ounces (oz)
Pounds (lb)	X	0.454	= Kilograms (kg)	X	2.205	= Pounds (lb)

Force
Ounces-force (ozf; oz)	X	0.278	= Newtons (N)	X	3.6	= Ounces-force (ozf; oz)
Pounds-force (lbf; lb)	X	4.448	= Newtons (N)	X	0.225	= Pounds-force (lbf; lb)
Newtons (N)	X	0.1	= Kilograms-force (kgf; kg)	X	9.81	= Newtons (N)

Pressure
Pounds-force per square inch (psi; lbf/in^2; lb/in^2)	X	0.070	= Kilograms-force per square centimetre (kgf/cm^2; kg/cm^2)	X	14.223	= Pounds-force per square inch (psi; lbf/in^2; lb/in^2)
Pounds-force per square inch (psi; lbf/in^2; lb/in^2)	X	0.068	= Atmospheres (atm)	X	14.696	= Pounds-force per square inch (psi; lbf/in^2; lb/in^2)
Pounds-force per square inch (psi; lbf/in^2; lb/in^2)	X	0.069	= Bars	X	14.5	= Pounds-force per square inch (psi; lbf/in^2; lb/in^2)
Pounds-force per square inch (psi; lbf/in^2; lb/in^2)	X	6.895	= Kilopascals (kPa)	X	0.145	= Pounds-force per square inch (psi; lbf/in^2; lb/in^2)
Kilopascals (kPa)	X	0.01	= Kilograms-force per square centimetre (kgf/cm^2; kg/cm^2)	X	98.1	= Kilopascals (kPa)
Millibar (mbar)	X	100	= Pascals (Pa)	X	0.01	= Millibar (mbar)
Millibar (mbar)	X	0.0145	= Pounds-force per square inch (psi; lbf/in^2; lb/in^2)	X	68.947	= Millibar (mbar)
Millibar (mbar)	X	0.75	= Millimetres of mercury (mmHg)	X	1.333	= Millibar (mbar)
Millibar (mbar)	X	0.401	= Inches of water (inH$_2$O)	X	2.491	= Millibar (mbar)
Millimetres of mercury (mmHg)	X	0.535	= Inches of water (inH$_2$O)	X	1.868	= Millimetres of mercury (mmHg)
Inches of water (inH$_2$O)	X	0.036	= Pounds-force per square inch (psi; lbf/in^2; lb/in^2)	X	27.68	= Inches of water (inH$_2$O)

Torque (moment of force)
Pounds-force inches (lbf in; lb in)	X	1.152	= Kilograms-force centimetre (kgf cm; kg cm)	X	0.868	= Pounds-force inches (lbf in; lb in)
Pounds-force inches (lbf in; lb in)	X	0.113	= Newton metres (Nm)	X	8.85	= Pounds-force inches (lbf in; lb in)
Pounds-force inches (lbf in; lb in)	X	0.083	= Pounds-force feet (lbf ft; lb ft)	X	12	= Pounds-force inches (lbf in; lb in)
Pounds-force feet (lbf ft; lb ft)	X	0.138	= Kilograms-force metres (kgf m; kg m)	X	7.233	= Pounds-force feet (lbf ft; lb ft)
Pounds-force feet (lbf ft; lb ft)	X	1.356	= Newton metres (Nm)	X	0.738	= Pounds-force feet (lbf ft; lb ft)
Newton metres (Nm)	X	0.102	= Kilograms-force metres (kgf m; kg m)	X	9.804	= Newton metres (Nm)

Power
Horsepower (hp)	X	745.7	= Watts (W)	X	0.0013	= Horsepower (hp)

Velocity (speed)
Miles per hour (miles/hr; mph)	X	1.609	= Kilometres per hour (km/hr; kph)	X	0.621	= Miles per hour (miles/hr; mph)

Fuel consumption
Miles per gallon, Imperial (mpg)	X	0.354	= Kilometres per litre (km/l)	X	2.825	= Miles per gallon, Imperial (mpg)
Miles per gallon, US (mpg)	X	0.425	= Kilometres per litre (km/l)	X	2.352	= Miles per gallon, US (mpg)

Temperature
Degrees Fahrenheit = (°C x 1.8) + 32

Degrees Celsius (Degrees Centigrade; °C) = (°F - 32) x 0.56

It is common practice to convert from miles per gallon (mpg) to litres/100 kilometres (l/100km), where mpg (Imperial) x l/100 km = 282 and mpg (US) x l/100 km = 235

Index

A

Air cleaner – 72, 106, 117, 118
Alternator – 91, 92
Antifreeze mixture – 68
Automatic transmission – 85

B

Battery – 90
Big-end bearing shells – 49
Braking system – 86

C

Cam followers – 51
Camshaft – 31
    drivebelt – 26, 57, 109
    end carrier – 51
    front oil seal – 35
Capacities, general – 5, 106
Clutch and transmission – 85
Cold start cable – 82
Connecting rods – 33, 49, 54
Conversion factors – 126
Cooling system – 63 *et seq*, 116
    antifreeze mixture – 68
    draining – 65
    fan – 67
    fault diagnosis – 69
    filling/flushing – 65
    radiator – 66
    thermostat – 66
    water pump – 68, 116
Core plugs – 51
Crankshaft – 48, 52
    oil seals – 36
    spigot bearing – 51
Cylinder bores – 49, 114
Cylinder head
    decarbonising – 48
    dismantling – 44
    gasket – 56
    overhaul – 47
    reassembly – 57
    removal and refitting – 29, 112

D

Driveplate (automatic transmission) – 51

E

Electrical system – 90 *et seq*
    alternator – 91, 92
    battery – 90
    starter motor – 96
    wiring diagrams, general – 103
Engine – 19 *et seq*, 109 *et seq*
    big-end bearing shells – 49
    camshaft – 26, 31, 51, 57, 109
    connecting rods – 33, 49, 54
    core plugs – 51
    crankshaft – 48, 51, 52
    cylinder bores – 49, 114
    cylinder head – 29, 44, 47, 48, 56, 57, 112
    dismantling – 41, 42
    driveplate (automatic transmission) – 51
    fault diagnosis – 17, 62
    firing order – 19
    flywheel – 35, 50, 55
    main bearings – 49, 52
    maintenance – 24, 108
    mountings – 36, 114
    oil change – 25
    oil pump – 33, 42, 54
    oil seals – 35, 36, 55
    oil types – 15
    pistons – 33, 48, 49, 54
        rings – 49, 54
    reassembly – 51, 52
    removal and refitting – 38, 39, 41, 59, 60, 115
    sump – 33, 54
    valves – 27, 51, 57, 111, 112
Exhaust gas recirculation (EGR) system – 124
Exhaust system – 83

F

Fan – 67
Fanbelt – 91
Fault diagnosis – 16 *et seq*
    cooling system – 69
    engine – 17, 62
    fuel system – 84

Index

Firing order 19
Flywheel 35, 50, 55
Fuel and exhaust systems 70 *et seq*, 116 *et seq*
    air cleaner 72, 117, 118
    cold start cable 82
    cold start device (vacuum-operated) 122
    exhaust gas recirculation (EGR) system 124
    fault diagnosis 83
    fuel filter 72, 106, 117
    fuel injection pump 74, 76, 121
    fuel injectors 78
    fuel tank 82
    glow plugs 80
    idle speed 73, 118 to 121
    idle stop solenoid 78
    maintenance 71, 116
    maximum speed 74, 118 to 121
    preheating system 79, 123
    throttle cable 80

G

Glow plugs 80

I

Idle speed 73, 118 to 121
Idle stop solenoid 78

L

Lubricants and fluids 15

M

Main bearing shells 49, 52
Maintenance, routine 11 *et seq*, 108, 109
    braking system 13
    clutch 13
    cooling system 13, 64, 65, 68
    electrical system 13, 90, 91
    engine 13, 24, 25, 26
    fuel system 13, 71, 72, 73, 80
    steering 13
    suspension 13
    transmission 13
    tyres/wheels 13, 88
Manifolds 83
Manual transmission 85
Maximum speed 74, 120
Mountings 36, 114

O

Oil
    change 25
    filter 22, 25
    pump 33, 42, 54
    seals 35, 36, 55

P

Pistons 33, 48, 49, 54
    rings 49, 54
Preheating system 79, 123

R

Radiator 66
Repair procedures, general 7
Routine maintenance see Maintenance

S

Safety first! 10
Spare parts 6, 16
Starter motor 96, 101
Sump 33, 54
Supplement: Revisions and information on later models 104
Suspension and steering 88 *et seq*

T

Thermostat 66
Throttle cable 80
Tools 8, 16
Transmission see Manual or Automatic transmission
Tyres 88, 108

V

Valve lifters 51, 57, 112
Valves 27, 111
Vacuum pump 86

W

Water pump 68, 116
Weights 5
Wheels 88
Wiring diagrams, general 103
Working facilities 9